Nonimaging Optics

OPTICAL SCIENCE AND ENGINEERING

Founding Editor **Brian J. Thompson**

Handbook of Optical and Laser Scanning, Second edition
Gerald F. Marshall, Glenn E. Stutz

Tunable Laser Applications, Second Edition
edited by F. J. Duarte

Laser Beam Shaping Applications, Second edition
Fred M. Dickey, Todd E. Lizotte

Lightwave Engineering
Yasuo Kokubun

Laser Safety: Tools and Training
edited by Ken Barat

Computational Methods for Electromagnetic and Optical Systems
John M. Jarem, Partha P. Banerjee

Biochemical Applications of Nonlinear Optical Spectroscopy
edited by Vladislav Yakovlev

Optical Methods of Measurement: Wholefield Techniques, Second Edition
Rajpal Sirohi

Fundamentals and Basic Optical Instruments
Daniel Malacara Hernández

Advanced Optical Instruments and Techniques
Daniel Malacara Hernández

Entropy and Information Optics: Connecting Information and Time, Second Edition
Francis T.S. Yu

Handbook of Optical Engineering, Second Edition, Two Volume Set
Daniel Malacara Hernández

Optical Materials and Applications
Moriaki Wakaki

Photonic Signal Processing: Techniques and Applications
Le Nguyen Binh

Practical Applications of Microresonators in Optics and Photonics
edited by Andrey B. Matsko

Near-Earth Laser Communications, Second Edition
edited by Hamid Hemmati

Nonimaging Optics: Solar and Illumination System Methods, Design, and Performance
Roland Winston, Lun Jiang, Vladimir Oliker

For more information about this series, please visit: https://www.routledge.com/Optical-Science
-and-Engineering/book-series/CRCOPTSCIENG

Nonimaging Optics
Solar and Illumination System Methods, Design, and Performance

Roland Winston

Lun Jiang

Vladimir Oliker

CRC Press
Taylor & Francis Group
Boca Raton London New York

CRC Press is an imprint of the
Taylor & Francis Group, an **informa** business

First published in 2021
by CRC Press
Taylor & Francis Group
6000 Broken Sound Parkway NW, Suite 300
Boca Raton, FL 33487-2742

First issued in paperback 2023

CRC Press is an imprint of Taylor & Francis Group, an Informa business
No claim to original U.S. Government works

ISBN 13: 978-1-032-65222-1 (pbk)
ISBN 13: 978-1-4665-8983-4 (hbk)
ISBN 13: 978-0-4291-6824-6 (ebk)

DOI: 10.1201/9780429168246

Library of Congress Cataloging-in-Publication Data

Names: Winston, Roland, author. | Jiang, Lun, author. | Oliker, Vladimir, 1945- author.
Title: Nonimaging optics : solar and illumination system methods, design, and performance / by Roland Winston, Lun Jiang, and Vladimir Oliker.
Description: First edition. | Boca Raton, FL : CRC Press/Taylor & Francis Group, [2020] | Series: Optical science and engineering | "Successor to Nonimaging Optics (Academic Press, 2005), by R. Winston, J. C. Miñano, and P. Benitez, with contributed chapters by N. Shatz and J. Bortz"--Preface. | Includes bibliographical references and index. | Summary: "This book provides a comprehensive look at the science, methods, designs, and limitations of nonimaging optics. It begins with an in-depth discussion on thermodynamically efficient optical designs and how they improve the performance and cost effectiveness of solar concentrating and illumination systems"-- Provided by publisher.
Identifiers: LCCN 2019049678 (print) | LCCN 2019049679 (ebook) | ISBN 9781466589834 (hardback ; acid-free paper) | ISBN 9780429168246 (ebook)
Subjects: LCSH: Solar collectors--Mathematical models. | Light--Transmission--Mathematical models. | Optical engineering--Mathematics. | Geometrical optics.
Classification: LCC TJ812 .W56 2005 (print) | LCC TJ812 (ebook) | DDC 621.47/2--dc23
LC record available at https://lccn.loc.gov/2019049678
LC ebook record available at https://lccn.loc.gov/2019049679

Visit the Taylor & Francis Web site at
https://www.taylorandfrancis.com

and the CRC Press Web site at
https://www.routledge.com

Typeset in Times
by Deanta Global Publishing Services

Contents

Preface

This book is the successor to *Nonimaging Optics* (Academic Press, 2005), by Roland Winston, J. C. Miñano, and P. Benitez, with chapters contributed by N. Shatz and J. Bortz. It in turn was preceded by *High Collection Nonimaging Optics* (Academic Press, 1989) and *Optics of Nonimaging Concentrators* (Academic Press, 1979), both by Walter T. Welford and Roland Winston. Walter T. Welford was one of the most distinguished optical scientists of his time. His work on aberration theory remains the definitive contribution to the subject. From 1976 until his untimely death in 1990, he took on the elucidation of nonimaging optics with the same characteristic vigor and enthusiasm he had applied to imaging optics. As a result, nonimaging optics developed from a set of heuristics to a complete subject. We continue to dedicate this book to his memory.

This series started as a project of my lifelong companion, the late Patricia Louise Winston, who upon meeting Walter T. Welford recognized the opportunity and value to future students of a monograph on nonimaging optics. She encouraged an application to the Guggenheim Foundation, which when successful led to a productive collaboration spanning 14 years.

This book incorporates much of the pre-1990 material as well as significant advances in the subject. These include elaborations of the flowline method designs for prescribed irradiance and a discussion of radiance that connects theory with measurement in a physical way. New chapters on Freeform Optics by Vladimir Oliker add significant insight, combining mathematical mapping techniques with classical nonimaging methods.

We acknowledge the significant effort by Sarah Boyd Jiang on the figures and editing of the text, also the efforts on the previous two books by Walter T. Welford and Roland Winston and the 2005 book by C. Miñano and P. Benitez. Lastly, we thank Robyn Lukens for her editorial help.

We will measure our success by the extent to which our readers advance the subject over the coming decade.

Roland Wilson
Lun Jiang
Vladimir Oliker

Photograph of Patricia Louise Winston around the time the first book in the series
was written

(courtesy of Roland Winston)

Photograph of Walter T. Welford

(courtesy of Jacqueline Welford)

Authors

Dr. Roland Winston is a leading figure in the field of nonimaging optics and its applications to solar energy. He is the inventor of the compound parabolic concentrator (CPC), used in solar energy, astronomy, and illumination. He is also a Guggenheim Fellow, a Franklin Institute medalist, past head of the University of Chicago Department of Physics, and a member of the founding faculty of University of California Merced, and he is currently the Director of UC Solar.

Dr. Lun Jiang is a Research Scientist at UC Solar. His expertise is with vacuum devices, nonimaging optics and solar thermal and hybrid systems, solar cooling, and solar desalination. In his Ph.D. thesis, he demonstrated two novel solar collectors that reach a working temperature above 200°C, without tracking. He led the receiver designing team for a vacuum hybrid receiver that generates both electricity and heat under 70x concentration, commissioned by Advance Research Project Agency-Energy (a segment of the U.S. Department of Energy).

Dr. Vladimir Oliker received his Ph.D. in mathematics from Leningrad University, USSR (the former Soviet Union). He has published over a hundred papers in the fields of pure and applied mathematics. Since 1984 he has been developing theoretical and computational methods for design of freeform optics. He is a Fulbright Fellow, senior member of SPIE, and is recipient of numerous research grants and awards.

1 Nonimaging Optical Systems and Their Uses

1.1 NONIMAGING COLLECTORS

Nonimaging concentrators and illuminators have several current and some potential applications. It is best to explain the general concept of a nonimaging concentrator by highlighting one of its applications: the utilization of solar energy. The radiation power density received from the sun at the earth's surface, often denoted by S, peaks at approximately 1 kW/m², depending on many factors. If we collect this power by absorbing it on a perfect blackbody, the equilibrium temperature T of the blackbody will be given by[*]

$$\sigma T^4 = S \tag{1.1}$$

where σ is the Stefan Boltzmann constant, 5.67×10^{-8} Wm⁻² K⁻⁴. In this example, the equilibrium temperature would be 364°K, or just below the boiling point of water. For many practical applications of solar energy this is sufficient, and many systems for domestic hot water heating based on this principle are available commercially for installation in private dwellings. However, for larger scale purposes or for generating electric power, a source of heat at 364°K has a low thermodynamic efficiency, since it is not practicable to get a very large temperature difference in whatever working fluid is being used in the heat engine. If we wanted, say, ≥300°C—a useful temperature for the generation of motive power—we would need to increase the power density S on the absorbing blackbody by a factor C of about 6 to 10 in Equation (1.1). This, briefly, is one use of a concentrator—to increase the power density of solar radiation. When it is stated plainly like that, the problem sounds trivial. The principles of the solution have been known since the days of Archimedes and his burning glass:[†] we simply have to focus the image of the sun with an image-forming system—a lens—and the result will be an increased power density. The problems to be solved are technical and practical, but they also lead to some interesting pure geometrical optics. The first question is that of the maximum concentration: how large a value of C is theoretically possible? The answer to this question is simple in all cases of interest. The next question—can the theoretical maximum concentration be achieved in practice?—is not as easy to answer. We shall see that there are limitations involving materials and manufacturing, as one would expect. But there are also limitations involving the kinds of optical systems that can actually be designed, as

---

[*] Ignoring various factors such as convection and conduction losses and radiation at lower effective emissivities.

[†] For an amusing argument concerning the authenticity of the story of Archimedes, see Stavroudis (1973).

opposed to those that are theoretically possible. This is analogous to the situation in classical lens design. The designers sometimes find that a certain specification cannot be fulfilled because it would require an impractically large number of refracting or reflecting surfaces. And sometimes they do not know whether it is in principle possible to achieve aberration corrections of a certain kind.

The natural approach of the classical optical physicist is to regard the problem as one of designing an image-forming optical system of a very large numerical aperture—that is, a small aperture ratio or *f*-number. One of the most interesting results to have emerged in this field is a class of very efficient concentrators that would have very large aberrations if they were used as image-forming systems. Nevertheless, as concentrators, they are substantially more efficient than image-forming systems and can be designed to meet or approach the theoretical limit. We shall call them *nonimaging concentrating collectors*, or *nonimaging concentrators* for short. Nonimaging is sometimes substituted by the word *anidolic* (from the Greek, meaning "without image") in languages such as Spanish and French because it's more specific. These systems are unlike any previously used optical systems. They have some of the properties of light pipes and some of the properties of image-forming optical systems but with very large aberrations. The development of the designs of these concentrators and the study of their properties have led to a range of new ideas and theorems in geometrical optics. In order to facilitate the development of these ideas, it is necessary to recapitulate some basic principles of geometrical optics, which is done in Chapter 2. In Chapter 3, we look at what can be done with conventional image-forming systems such as concentrators, and we show how they necessarily fall short of ideal performance. In Chapter 4, we describe one of the basic nonimaging concentrators, the compound parabolic concentrator, and we obtain its optical properties. Chapter 5 is devoted to several developments of the basic compound parabolic concentrator: with plane absorber, mainly aimed at decreasing the overall length; with nonplane absorber; and with generalized edge-ray wave fronts, which is the origin of the tailored designs. In Chapter 6, we examine in detail the flowline approach to nonimaging. Chapter 7 will be about the exposition of the emerging field of freeform optics. Chapter 8 is about the wave description of the optical measurements.

1.2 FAILURE OF IMAGING OPTICS

An example of the failure of imaging optics to confront real-world situations is shown in Figure 1.1. The point object A is at the center of a spherical reflecting cavity and is also one focus of an elliptical reflecting cavity. The point object B is at the other focus. If we start A and B at the same temperature, the probability of radiation from B reaching A is clearly higher than A reaching B. So we conclude that A warms up while B cools off, in violation of the second law of thermodynamics. The paradox is resolved by making A and B extended objects, no matter how small. In fact, a physical object with temperature has many degrees of freedom and cannot be point-like. Then, the correct cavity is not elliptical, but a nonimaging shape that ensures an efficient equal radiation transfer between A and B.* It is worth mentioning that the

---

* *The Ellipsoid Paradox in Thermodynamics,* W. T. Welford and R. Winston, *Journal of Statistical Physics,* Vol. 28, No 3, 1982

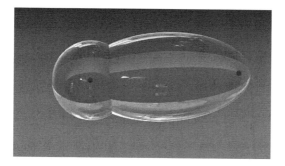

FIGURE 1.1 The ellipse paradox: the ellipse images "point" object B (right) at "point" object A (left) "perfectly" and the sphere images A on itself "perfectly."

correct nonimaging design does not converge into the ellipse/sphere configuration at the limit, i.e. zero, toward which the size of A and B tend.

1.3 DEFINITION OF THE CONCENTRATION RATIO; THE THEORETICAL MAXIMUM

From the simple argument in Section 1.1 we see that the most important property of a concentrator is the ratio of area of input beam divided by area of output beam; this is because the equilibrium temperature of the absorbing body is proportional to the fourth root of this ratio.

We have denoted this ratio by C and called it the concentration ratio. Initially, we modeled a concentrator as a box with a plane entrance aperture of area A and a plane exit aperture of area A' that is just large enough to allow all transmitted rays to emerge (see Figure 1.3). Then the concentration ratio is

$$C = A / A' \qquad (1.2)$$

Consider the following problem described in Figure 1.2, a design that maximizes the temperature of the absorber T_3, while maintaining the source at T_1 (a heat reservoir,

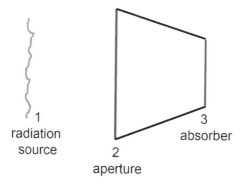

FIGURE 1.2 A schematic of the general concentrator problem.

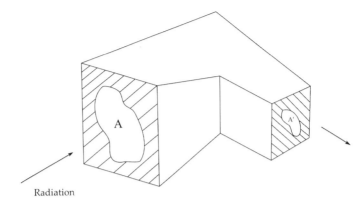

Radiation

FIGURE 1.3 Schematic diagram of a concentrator. The input and output surfaces can face in any direction; they are drawn in the figure so both can be seen. It is assumed that the aperture A' is just large enough to permit all rays passed by the internal optics that have entered within the specified collecting angle to emerge.

like the sun), and introducing probabilities P_{ij} where P_{13} is the probability of radiation from wall 1 reaching wall 3, and so on. The condition is $P_{31} = 1$. In cases where P_{33} is not 0 (as in a cavity receiver) the generalization is $P_{31} = 1 - P_{33}$ which still maximizes P_{31}. These conditions are intuitively self-evident. Here we sketch a proof: if $P_{31} < 1$, then some energy, say P_{3x}, is exchanged with a body x which is, in general, at a lower temperature than the source 1. Therefore, at equilibrium, T_3 will reach a temperature intermediate between T_1 and T_x. Stated the other way, if T_3 reaches T_1 at equilibrium, then all the energy exchange must be between A_1 and A_3 since the other objects will in general be at a lower temperature. Additionally, for an efficient concentrating system ($P_{12} = P_{13}$) the concentration value is $1/P_{21}$. In the case of photovoltaic systems, the condition states that light starting from the solar cell would invariably reach the source. The application of this concept should lead to the development of more thermodynamically efficient thermal designs as well as higher efficiency solar cells.

In other words, the second law guarantees:

$$A_1 P_{12} = A_2 P_{21}$$

$$A_1 P_{13} = A_3 P_{31}$$

In a lossless system, the first law or energy conservation guarantees:

$$A_1 P_{13} = A_1 P_{12}$$

Therefore,

$$A_2 P_{21} = A_3 P_{31}$$

$$C = \frac{A_2}{A_3} = \frac{P_{31}}{P_{21}} \leq \frac{1}{P_{21}}$$

In the preceding definition, it was tacitly assumed that compression of the input beam occurred in both the dimensions transverse to the beam direction, as in ordinary lens systems. In solar energy technology there is a large class of systems in which the beam is compressed in only one dimension. In such systems all the operative surfaces, reflecting and refracting, are cylindrical with parallel generators (but not in general circular cylindrical). Thus, a typical shape would be a parabolic trough, with the absorbing body lying along the trough. Such long trough collectors have the obvious advantage of not needing to be guided to follow the daily movement of the sun across the sky. The two types of concentrator are sometimes called three- and two-dimensional, or 3D and 2D, concentrators. The names 3D and 2D are also used in this book to denote that the optical device has been designed in 3D geometry or in 2D geometry (in the latter case, the real concentrator, which of course exists in a 3D space, is obtained by rotational or translational symmetry from the 2D design). In these cases we will use the name 2D design or 3D design to differentiate from a 2D or a 3D concentrator. The 2D concentrators are also called linear concentrators. The concentration ratio of a linear concentrator is usually given as the ratio of the transverse input and output dimensions, measured perpendicular to the straight-line generators of the trough. The question immediately arises whether there is any upper limit to the value of C, and we shall see that there is. The result, proved later, is very simple for the 2D case and for the 3D case with an axis of revolution symmetry (rotational concentrator). Suppose the input and output media both have a refractive index of unity, and let the incoming radiation be from a circular source at infinity subtend a semiangle θ_i. Then the theoretical maximum concentration in a rotational concentrator is

$$C_{max} = 1 / \sin^2 \theta_i \tag{1.3}$$

$$\int d\Omega \cos\theta = \int \sin\theta \, d\theta \, d\psi \cos\theta$$

$$= \int_0^\theta \cos\theta \, d(\cos\theta) \int_0^{2\pi} d\psi$$

$$= 2\pi \frac{1}{2} \left(1 - \cos^2\theta\right)$$

$$= \pi \sin^2\theta$$

$$P_{21} = \frac{\pi \sin^2\theta}{\pi \sin^2 \frac{\pi}{2}} = \sin^2\theta$$

Under this condition the rays emerge at all angles up to $\pi/2$ from the normal to the exit face, as shown in Figure 1.3. For a linear concentrator the corresponding value will be $1/\sin\theta_i$. The next question that arises is, can actual concentrators be designed with the theoretically best performance? In asking this question we make

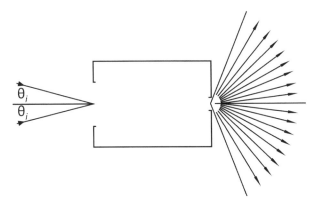

FIGURE 1.4 Incident and emergent ray paths for an ideal 3D concentrator with symmetry about an axis of revolution. The exit aperture diameter is $\sin\theta_i$ times the exit aperture diameter; the rays emerge from all points in the exit aperture over a solid angle $2/\pi$.

certain idealizing assumptions—for example, that all reflecting surfaces have 100% reflectivity, that all refracting surfaces can be perfectly antireflection-coated, that all shapes can be made exactly right, and so forth. We shall then see that the following answers are obtained: (1) 2D concentrators can be designed with the theoretical maximum concentration; (2) 3D concentrators can also have the theoretical maximum concentration if they use variable refractive index material or a pile of infinitely thin surface waveguides, properly shaped; and (3) some rotational symmetric concentrators can have the theoretical maximum concentration. In case (3) it appears for other types of design that it is possible to approach indefinitely close to the theoretical maximum concentration either by sufficiently increasing the complexity of the design or by incorporating materials that are in principle possible but in practice not available. For example, we might specify a material of very high refractive index—say, 5—although this is not actually available without large absorption in the visible part of the spectrum.

1.4   USES OF CONCENTRATORS

This application of concentrators as solar energy amplifiers has, of course, stimulated great developments in the design and fabrication of concentrators. But it is by no means the only application. The particular kind of nonimaging concentrator that has given rise to these great developments was originally conceived as a device for collecting as much light as possible from a luminous volume (the gas or fluid of a Cerenkov counter) over a certain range of solid angles and sending it onto the cathode of a photomultiplier. Since photomultipliers are limited in size and since the volume in question was of order 1 m3, this is clearly a concentrator problem (Hinterberger & Winston, 1966a,b).

Subsequently the concept was applied to infrared detection (Harper et al., 1976), where it is well known that the noise in the system for a given type of detector increases with the surface area of the detector (other things being equal).

1.5   USES OF ILLUMINATORS

Nonimaging collectors are also used in illumination. The source (a filament, an LED, etc.) in general emits in a wide angular spread at low intensity, and the problem consists of designing an optical device that efficiently collimates this radiation so it is emitted in a certain angular emitting region, which is smaller than the angular emitting region of the source. The problem is conceptually similar to the concentrating problem, substituting aperture areas for angular region sizes. We will see soon that both statements are equivalent (Figure 1.4).

REFERENCES

Harper, D. A., Hildebrand, R. H., Pernic, R., and Platt, R. (1976). Heat trap: An optimised far infrared field optics system. *Appl. Opt.* **15**, 53–60.

Hinterberger, H., and Winston, R. (1966a). Efficient light coupler for threshold Cerenkov counters. *Rev. Sci. Instrum.* **37**, 1094–1095.

Hinterberger, H., and Winston, R. (1966a). Gas Cerenkov counter with optimized light-collection efficiency. *Proc. Int. Conf. Instrum. High Energy Phys.* 205–206.

Winston, R., and Enoch, J. M. (1971). Retinal cone receptor as an ideal light collector. *J. Opt. Soc. Am.* **61**, 1120–1122.

2 Some Basic Ideas in Geometrical Optics

2.1 THE CONCEPTS OF GEOMETRICAL OPTICS

Geometrical optics is used as the basic tool in designing almost any optical system, image-forming or not. We use the intuitive ideas of a ray of light, roughly defined as the path along which light energy travels, together with surfaces that reflect or transmit the light. When light is reflected from a smooth surface, it obeys the well-known law of reflection, which states that the incident and reflected rays make equal angles with the normal to the surface and that both rays and the normal lie in one plane. When light is transmitted, the ray direction is changed according to the law of refraction: Snell's law. This law states that the sine of the angle between the normal and the incident ray bears a constant ratio to the sine of the angle between the normal and the refracted ray; again, all three directions are coplanar.

A major part of the design and analysis of concentrators involves ray-tracing—that is, following the paths of rays through a system of reflecting and refracting surfaces. This is a well-known process in conventional lens design, but the requirements are somewhat different for concentrators, so it will be convenient to state and develop the methods ab initio. This is because in conventional lens design the reflecting or refracting surfaces involved are almost always portions of spheres, and the centers of the spheres lie on one straight line (axisymmetric optical system) so that special methods that take advantage of the simplicity of the forms of the surfaces and the symmetry can be used. Nonimaging concentrators do not, in general, have spherical surfaces. In fact, sometimes there is no explicitly analytical form for the surfaces, although usually there is an axis or a plane of symmetry. We shall find it most convenient, therefore, to develop ray-tracing schemes based on vector formulations but with the details covered in computer programs on an ad hoc basis for each different shape.

In geometrical optics we represent the power density across a surface by the density of ray intersections with the surface and the total power by the number of rays. This notion, reminiscent of the useful but outmoded "lines of force" in electrostatics, works as follows. We take N rays spaced uniformly over the entrance aperture of a concentrator at an angle of incidence θ, as shown in Figure 2.1.

Suppose that after tracing the rays through the system only N' emerge through the exit aperture, the dimensions of the latter being determined by the desired concentration ratio. The remaining $N - N'$ rays are lost by processes that will become clear when we consider some examples. Then the power transmission for the angle θ is taken as N'/N. This can be extended to cover a range of angle θ as required. Clearly, N must be taken largely enough to ensure that a thorough exploration of possible ray paths in the concentrator is made.

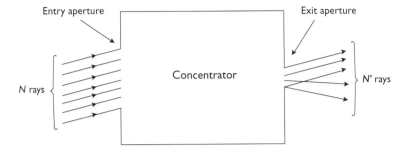

FIGURE 2.1 Determining the transmission of a concentrator by ray-tracing.

2.2 FORMULATION OF THE RAY-TRACING

To formulate a ray-tracing procedure suitable for all cases, it is convenient to put the laws of reflection and refraction into vector form. Figure 2.2 shows the geometry with unit vectors r and r'' along the incident and reflected rays and a unit vector n along the normal pointing into the reflecting surface. To see this, we know that n bisects $-r''$ and r, so that $r'' - r$ is parallel to n, say $a\,n$, to find a, we dot both sides by n, and a follows.

$$r'' = r - 2(n \cdot r)n \qquad (2.1)$$

Thus, to ray-trace "through" a reflecting surface, first we have to find the point of incidence, a problem of geometry involving the direction of the incoming ray and the known shape of the surface. Then we have to find the normal at the point of incidence—again a problem of geometry. Finally, we have to apply Equation (2.1) to find the direction of the reflected ray. The process is then repeated if another reflection is

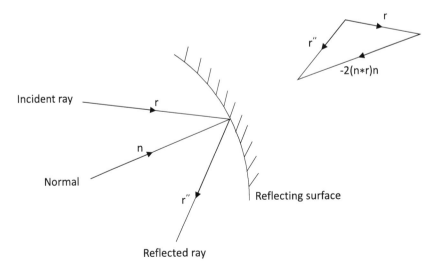

FIGURE 2.2 Vector formulation of reflection. r, r'' and n are all unit vectors.

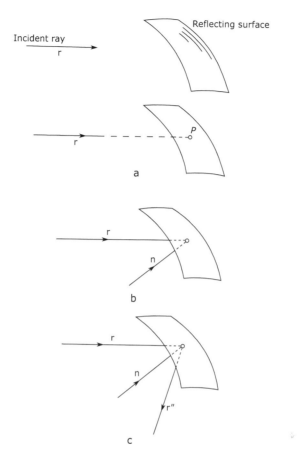

FIGURE 2.3   The stages in ray-tracing a reflection. (a) Find the point of incidence P. (b) Find the normal at P. (c) Apply Equation (2.1) to find the reflected ray $\mathbf{r}\leq$.

to be taken into account. These stages are illustrated in Figure 2.3. Naturally, in the numerical computation the unit vectors are represented by their components—that is, the direction cosines of the ray or normal with respect to some Cartesian coordinate system used to define the shape of the reflecting surface. Ray-tracing through a refracting surface is similar, but first we have to formulate the law of refraction vectorially. Figure 2.4 shows the relevant unit vectors. It is similar to Figure 2.2 except that $\mathbf{r}'$ is a unit vector along the refracted ray.

We denote by n, n' the refractive indexes of the media on either side of the refracting boundary; the refractive index is a parameter of a transparent medium to the speed of light in the medium. Specifically, if c is the speed of light in a vacuum, the speed in a transparent material medium is c/n, where n is the refractive index. For visible light, values of n range from unity to about 3 for usable materials in the visible spectrum. The law of refraction is usually stated in the form

$$n'\sin I' = n\sin I \tag{2.2}$$

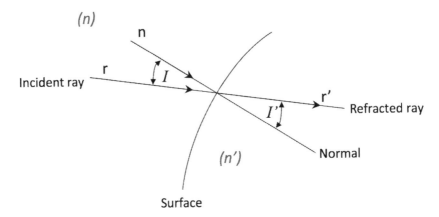

FIGURE 2.4   Vector formulation of refraction.

where I and I' are the angles of incidence and refraction, as in the figure, and where the coplanarity of the rays and the normal is understood. The vector formulation

$$n'\boldsymbol{r}' \times \boldsymbol{n} = n\,\boldsymbol{r} \times \boldsymbol{n} \tag{2.3}$$

contains everything, since the modulus of a vector product of two unit vectors is the sine of the angle between them. This can be put in the form most useful for ray-tracing by multiplying through vectorially by n to give

$$n'\boldsymbol{r}' = n\boldsymbol{r} + \left(n'\boldsymbol{r}' \cdot \boldsymbol{n} - n\boldsymbol{r} \cdot \boldsymbol{n}\right)\boldsymbol{n} \tag{2.4}$$

which is the preferred form for ray-tracing.* The complete procedure then parallels that for reflection explained by means of Figure 2.3. We find the point of incidence, then the direction of the normal, and finally the direction of the refracted ray. Details of the application to lens systems are given, for example, by Welford (1974, 1986).

If a ray travels from a medium of refractive index n toward a boundary with another of index $n' < n$, then as can be seen from Equation (2.2) it would be possible to have *sin I'* greater than unity. Under this condition it is found that the ray is completely reflected at the boundary. This is called total internal reflection, and we shall find it a useful effect in concentrator design.

2.3   ELEMENTARY PROPERTIES OF IMAGE-FORMING OPTICAL SYSTEMS

In principle, the use of ray-tracing tells us all there is to know about the geometrical optics of a given optical system, image-forming or not. However, ray-tracing alone is

---

* The method of using Equation (2.4) numerically is not so obvious as it is for Equation (2.2), since the coefficient of $\boldsymbol{n}$ in Eq (2.4) is actually $n'\cos I' - n\cos I$. Thus, it might appear that we have to find $\boldsymbol{r}'$ before we can use the equation. The procedure is to find cos I' via Equation (2.2) first, and then Equation (2.4) is needed to give the complete three-dimensional picture of the refracted ray.

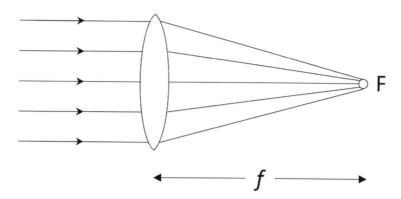

FIGURE 2.5 A thin converging lens bringing parallel rays to a focus. Since the lens is technically "thin," we do not have to specify the exact plane in the lens from which the focal length f is measured.

of little use for inventing new systems having properties suitable for a given purpose. We need to have ways of describing the properties of optical systems in terms of general performance, such as, for example, the concentration ratio C introduced in Chapter 1. In this section we shall introduce some of these concepts. Consider first a thin converging lens such as one that would be used as a magnifier or in eyeglasses for a farsighted person (see Figure 2.5). By "thin" we mean that its thickness can be neglected for the purposes of our discussion. Elementary experiments show us that if we have rays coming from a point at a great distance to the left, so that they are substantially parallel as in the figure, the rays meet approximately at a point F, the focus. The distance from the lens to F is called the focal length, denoted by f. Elementary experiments also show that if the rays come from an object of finite size at a great distance, the rays from each point on the object converge to a separate focal point, and we get an image. This is, of course, what happens when a burning glass forms an image of the sun or when the lens in a camera forms an image on film. This is indicated in Figure 2.6, where the object subtends the (small) angle 2q. It is then found that the size of the image is $2f\theta$. This is easily seen by considering the rays through the center of the lens, since these pass through undeviated. Figure 2.6 contains one of the fundamental concepts we use in concentrator theory, the concept of a beam of light of a certain diameter and angular extent.

The diameter is that of the lens—say, $2a$—and the angular extent is given by 2θ. These two can be combined as a product, usually without the factor 4, giving θa, a quantity known by various names including extent, étendue, acceptance, and Lagrange invariant. It is, in fact, an invariant through the optical system, provided that there are no obstructions in the light beam and provided we ignore certain losses due to properties of the materials, such as absorption and scattering.

For example, at the plane of the image the étendue becomes the image height θf multiplied by the convergence angle a/f of the image-forming rays, giving again θa. In discussing 3D systems—for example, an ordinary lens such as we have supposed Figure 2.6 to represent—it is convenient to deal with the square of this quantity, $a^2\theta^2$. This is also sometimes called the étendue, but generally it is clear from the

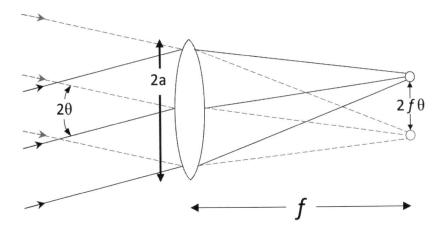

FIGURE 2.6 An object at infinity has an angular subtense 2θ. A lens of focal length f forms an image of size $2f\theta$.

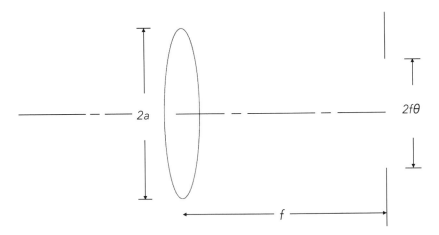

FIGURE 2.7 An optical system of acceptance, throughput, or étendue $a2q2$.

context and from dimensional considerations which form is intended. The 3D form has an interpretation that is fundamental to the theme of this book. Suppose we put an aperture of diameter $2f\theta$ at the focus of the lens, as in Figure 2.7. Then this system will only accept rays within the angular range $\pm\theta$ and inside the diameter $2a$. Now suppose a flux of radiation B (in $Wm^{-2}\ sr^{-1}$) is incident on the lens from the left.* The system will actually accept a total flux $(Bp^2\theta^2a^2)W$; thus, the étendue or acceptance θ^2a^2 is a measure of the power flow that can pass through the system.

The same discussion shows how the concentration ratio C appears in the context of classical optics. The accepted power $(Bp^2\theta^2a^2)W$ must flow out of the aperture to the right of the system, if our preceding assumptions about how the lens forms

---

* In full, B watts per square meter per steradian solid angle.

an image are correct* and if the aperture has the diameter $2f\theta$. Thus, our system is acting as a concentrator with concentration ratio $C = \left(\dfrac{2a}{2f\theta}\right)^2 = \left(\dfrac{a}{f\theta}\right)^2$ for the input semiangle θ.

Let us relate these ideas to practical cases. For solar energy collection we have a source at infinity that subtends a semiangle of approximately 0.005 rad (1/4°) so that this is the given value of q, the collection angle. Clearly, for a given diameter of lens we gain concentration by reducing the focal length as much as possible.

2.4 ABERRATIONS IN IMAGE-FORMING OPTICAL SYSTEMS

According to the simplified picture presented in Section 2.3, there is no reason why we could not make a lens system with an indefinitely large concentration ratio by simply decreasing the focal length sufficiently. This is, of course, not so, partly because of aberrations in the optical system and partly because of the fundamental limit on concentration stated in Section 1.2.

We can explain the concept of aberrations by looking again at our example of the thin lens in Figure 2.5. We suggested that the parallel rays all converged after passing through the lens to a single point F. In fact, this is only true in the limiting case when the diameter of the lens is taken as indefinitely small. The theory of optical systems under this condition is called paraxial optics or Gaussian optics, and it is a very useful approximation for getting at the main large-scale properties of image-forming systems. If we take a simple lens with a diameter that is a sizable fraction of the focal length—say, $f/4$—we find that the rays from a single point object do not all converge to a single image point. We can show this by ray-tracing. We first set up a proposed lens design, as shown in Figure 2.8. The lens has curvatures (reciprocals of radii) c_1 and c_2, center thickness d, and refractive index n. If we neglect the central thickness for the moment, then it is shown in specialized treatment (e.g., Welford, 1986) that the focal length f is given in paraxial approximation by

$$1/f = (n-1)(c_1 - c_2) \tag{2.5}$$

and we can use this to get the system to have roughly the required paraxial properties.

Now we can trace rays through the system as specified, using the method outlined in Section 2.2 (details of ray-tracing methods for ordinary lens systems are given in, for example, Welford, 1974). These will be exact or finite rays, as opposed to paraxial rays, which are implicit in the Gaussian optics approximation. The results for the lens in Figure 2.8 would look like Figure 2.9. This shows rays traced from an object point on the axis at infinity—that is, rays parallel to the axis.

In general, for a convex lens the rays from the outer part of the lens aperture meet the axis closer to the lens than the paraxial rays do. This effect is known as spherical aberration. (The term is misleading, since the aberration can occur in systems with nonspherical refracting surfaces, but there seems little point in trying to change it at the present advanced state of the subject.)

---

* As we shall see, these assumptions are only valid for limitingly small apertures and objects.

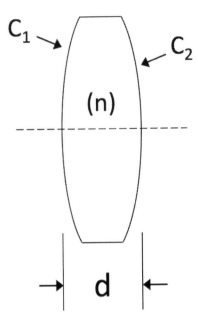

FIGURE 2.8   Specification of a single lens. The curvature $c1$ is positive as shown, and $c2$ is negative.

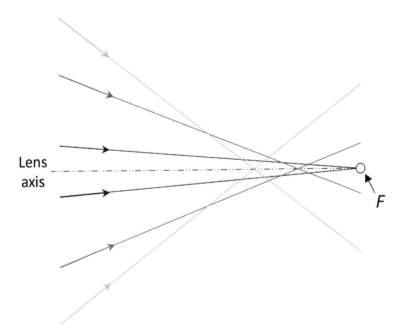

FIGURE 2.9   Rays near the focus of a lens showing spherical aberration.

Spherical aberration is perhaps the simplest of the different aberration types to describe, but it is just one of many. Even if we were to choose the shapes of the lens surfaces so as to eliminate the spherical aberration or were to eliminate it in some other way, we would still find that the rays from object points away from the axis do not form point images—in other words, there would be oblique or off-axis aberrations.

Also, the refractive index of any material medium changes with the wavelength of the light, and this produces chromatic aberrations of various kinds. We do not at this stage need to go into the classification of aberrations very deeply, but this preliminary sketch is necessary to show the relevance of aberrations to the attainable concentration ratio.

2.5 THE EFFECT OF ABERRATIONS IN AN IMAGE-FORMING SYSTEM ON THE CONCENTRATION RATIO

Questions regarding the extent to which it is theoretically possible to eliminate aberrations from an image-forming system have not yet been fully answered. In this book we shall attempt to give answers adequate for our purposes, although they may not be what the classical lens designers want. For the moment, let us accept that it is possible to eliminate spherical aberration completely, but not the off-axis aberrations, and let us suppose that this has been done for the simple collector of Figure 2.7. The effect will be that some rays of the beam at the extreme angle θ will fall outside the defining aperture of diameter $2f\theta$. We can see this more clearly by representing an aberration by means of a spot diagram. This is a diagram in the image plane with points plotted to represent the intersections of the various rays in the incoming beam. Such a spot diagram for the extreme angle θ might appear as in Figure 2.10. The ray through the center of the lens (the principal ray in lens theory) meets the rim of the collecting aperture by definition, and thus a considerable amount of the flux does not get through. Conversely, it can be seen (in this case at least) that some flux from beams at a larger angle than θ will be collected.

We display this information on a graph such as in Figure 2.11. This shows the proportion of light collected at different angles up to the theoretical maximum, θ_{max}. An ideal collector would behave according to the full line—that is, it would collect all light flux within θ_{max} and none outside. At this point it may be objected that all we need to do to achieve the first requirement is to enlarge the collecting aperture slightly, and the second requirement does not matter. However, we recall that our aim is to achieve maximum concentration because of the requirement for high-operating temperature so that the collector aperture must not be enlarged beyond $2f\theta$ diameter.

Frequently in discussions of aberrations in books on geometrical optics, the impression is given that aberrations are in some sense "small." This is true in optical systems designed and made to form reasonably good images, such as camera lenses. But these systems do not operate with large enough convergence angles (a/f in the notation for Figure 2.6) to approach the maximum theoretical concentration ratio. If we were to try to use a conventional image-forming system under such conditions, we would find that the aberrations would be very large and that they would severely depress the concentration ratio. Roughly, we can say that this is one limitation that

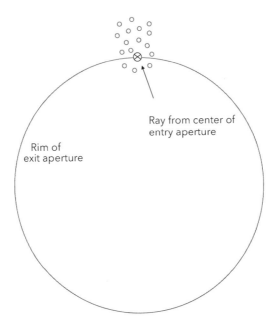

FIGURE 2.10 A spot diagram for rays from the beam at the maximum entry angle for an image-forming concentrator. Some rays miss the edge of the exit aperture due to aberrations, and the concentration is thus less than the theoretical maximum.

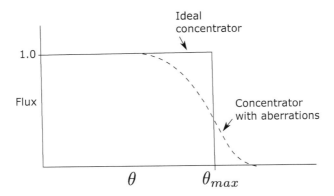

FIGURE 2.11 A plot of collection efficiency against angle. The ordinate is the proportion of flux entering the collector aperture at angle q that emerges from the exit aperture.

has led to the development of the new, nonimaging concentrators. Nevertheless, we cannot say that imaging-forming is incompatible with attaining maximum concentration. We will show, later, examples in which both properties are combined.

2.6   THE OPTICAL PATH LENGTH AND FERMAT'S PRINCIPLE

There is another way of looking at geometrical optics and the performance of optical systems, which we also need to outline for the purposes of this book. We noted

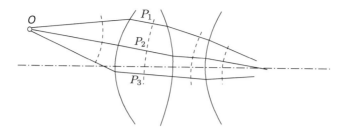

FIGURE 2.12   Rays and (in broken line) geometrical wave fronts.

FIGURE 2.13   Fermat's principle. It is assumed in the diagram that the medium has a continuously varying refractive index. The solid line path has a stationary optical path length from A to B and is therefore a physically possible ray path.

in Section 2.2 that the speed of light in a medium of refractive index n is c/n, where c is the speed in a vacuum. Thus, light travels a distance s in the medium in time $s/v = ns/c$; that is, the time taken to travel a distance s in a medium of refractive index n is proportional to ns. The quantity ns is called the optical path length corresponding to the length s. Suppose we have a point source O emitting light into an optical system, as in Figure 2.12. We can trace any number of rays through the system, as outlined in Section 2.2, and then we can mark off along these rays points that are all at the same optical path length from O—say, $P1$, $P2$... We do this by making the sum of the optical path lengths from O in each medium the same—that is,

$$\Sigma ns = \text{const.} \tag{2.6}$$

in an obvious notation. These points can be joined to form a surface (we are supposing rays out of the plane of the diagram to be included), which would be a surface of constant phase of the light waves if we were thinking in terms of the wave theory of light.* We call it a geometrical wave front, or simply a wave front, and we can construct wave fronts at all distances along the bundle of rays from O.

We now introduce a principle that is not as intuitive as the laws of reflection and refraction but that leads to results that are indispensable to the development of the theme of this book. It is based on the concept of optical path length, and it is a way of predicting the path of a ray through an optical medium. Suppose we have any optical medium that can have lenses and mirrors and can even have regions of continuously varying refractive index. We want to predict the path of a light ray between two points in this medium—say, A and B in Figure 2.13. We can propose an infinite number of possible paths, of which three are indicated. But unless A and B happen to be object and image—and we assume they are not—only one or perhaps a small finite

---

* This construction does not give a surface of constant phase near a focus or near an edge of an opaque obstacle, but this does not affect the present applications.

number of paths will be physically possible—in other words, paths that rays of light could take according to the laws of geometrical optics. Fermat's principle in the form most commonly used states that a physically possible ray path is one for which the optical path length from A to B is an extremum as compared to neighboring paths. For "extremum" we can often write "minimum," as in Fermat's original statement. It is possible to derive all of geometrical optics—that is, the laws of refraction and reflection—from Fermat's principle. It also leads to the result that the geometrical wave fronts are orthogonal to the rays (the theorem of Malus and Dupin); that is, the rays are normal to the wave fronts. This in turn tells us that if there is no aberration—if all rays meet at one point—then the wave fronts must be portions of spheres. So if there is no aberration, the optical path length from object point to image point is the same along all rays. Thus, we arrive at an alternative way of expressing aberrations: in terms of the departure of wave fronts from the ideal spherical shape. This concept will be useful when we come to discuss the different senses in which an image-forming system can form "perfect" images.

2.7   THE OPTICAL LAGRANGIAN

Geometrical optics can be stated in the form

$$\delta \int_{P_1}^{P_2} n(x,y,z)\,ds = 0 \tag{2.7}$$

where ds is an element of the ray path from $P1$ to $P2$. This can be written in the form

$$\delta \int_{P_1}^{P_2} \mathcal{L}(x,y,z,\dot{x},\dot{y})\,dz = 0 \tag{2.8}$$

where

$$\mathcal{L}(x,y,z,\dot{x},\dot{y}) = n(x,y,z)\sqrt{1+\dot{x}^2+\dot{y}^2} \tag{2.9}$$

and the dots denote differentiation with respect to z. Also, we define

$$p = \frac{n\dot{x}}{\sqrt{1+\dot{x}^2+\dot{y}^2}}, \quad q = \frac{n\dot{y}}{\sqrt{1+\dot{x}^2+\dot{y}^2}} \tag{2.10}$$

The analogy is to regard L as the Lagrangian function of a mechanical system in which x and y are two generalized coordinates, p and q are the corresponding generalized momenta, and z corresponds to the time axis. On this basis the ordinary development of mechanics can be carried out, such as by Luneburg (1964, Article 18), by solving the variational problem of Equation (2.7). The Hamiltonian is found to have the value

$$\mathcal{H} = -\sqrt{n^2 - p^2 - q^2} \tag{2.11}$$

In other words, it is $-nN$ where N is the z-direction cosine, and, of course, p and q as just defined are respectively equal to nL and nM. The phase space for this system has the four coordinates (x, y, p, q), and Liouville's theorem in statistical mechanics can be invoked immediately to state that phase space volume is conserved (Goldstein, classical mechanics).

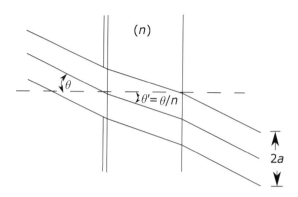

FIGURE 2.14   Inside a medium of refractive index n the étendue becomes $n^2a^2\theta^2$.

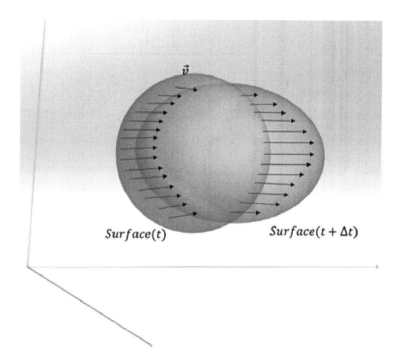

FIGURE 2.15   The phase space volume remains constant.

2.8 THE FERMI VECTOR METHOD

"An important result deserves more than one proof."

Consider the Legendre transformation of the Optical Lagrangian:

Construct $p_x \dot{x} + p_y \dot{y} - \mathcal{L}(x, y, \dot{x}, \dot{y}) = \mathcal{H}$

$$d\mathcal{H} = p_x d\dot{x} + dp_x \dot{x} + p_y d\dot{y} + dp_y \dot{y}$$

$$- \left[\frac{\partial \mathcal{L}}{\partial x} dx + \frac{\partial \mathcal{L}}{\partial y} dy + \frac{\partial \mathcal{L}}{\partial \dot{x}} d\dot{x} + \frac{\partial \mathcal{L}}{\partial \dot{y}} d\dot{y} \right]$$

$$= -\dot{p}_x dx - \dot{p}_y dy + \dot{x} dp_x + \dot{y} dp_y$$

This results in $\dfrac{\partial H}{\partial x} = -\dot{p}_x, \dfrac{\partial H}{\partial y} = -\dot{p}_y, \dfrac{\partial H}{\partial p_x} = \dot{x}, \dfrac{\partial H}{\partial p_y} = \dot{y}$, which is similar to

Hamiltonian equations.

Notice that $\dfrac{\partial \dot{p}_x}{\partial p_x} = -\dfrac{\partial}{\partial p_x}\left(\dfrac{\partial H}{\partial x} \right) = -\dfrac{\partial}{\partial x}\left(\dfrac{\partial H}{\partial p_x} \right) = \dfrac{\partial \dot{x}}{\partial x}$, etc.

Now construct a vector $\vec{W} = (\dot{x}, \dot{y}, \dot{p}_x, \dot{p}_y)$, and notice that

$\nabla \cdot \vec{W} = \dfrac{\partial W}{\partial \dot{x}} + \dfrac{\partial W}{\partial \dot{y}} + \dfrac{\partial W}{\partial \dot{p}_x} + \dfrac{\partial W}{\partial \dot{p}_y} = \dfrac{\partial \dot{p}_x}{\partial p_x} + \dfrac{\partial \dot{x}}{\partial x} + \dfrac{\partial \dot{p}_y}{\partial p_y} + \dfrac{\partial \dot{y}}{\partial y} = 0$. This means the field of

four-dimensional vector $\vec{W}$ has the important property of divergence being zero. In other words, the four-dimensional hyperspace of (x, y, p_x, p_y) has the property of conservation of volume as all the light rays evolve in an optical system.

Here we will also offer an intuitive analogy using fluid (Figure 2.15).

Let $\mathcal{V}$ be the volume surrounded by a closed *Surface* in a field with small integral elements denoted as $d\vec{s}$, where $\vec{v}$ is the speed of small particles of an imaginary fluid. As the fluid starts to flow according to the change of t, and the *Surface*(t) starts to evolve into shape as *Surface*$(t + \Delta t)$, the enclosed volume of $\mathcal{V}(t + \Delta t)$ will also change as:

$$\mathcal{V}(t + \Delta t) - \mathcal{V}(t) = \oiint_{Surface(t)} \vec{v} \Delta t \cdot d\vec{s}$$

Where $d\vec{s}$ is the surface vector pointing along the normal direction outwards.

Using Gauss's theorem:

$$\oiint_{Surface(t)} \vec{v} \Delta t \cdot d\vec{s} = \iiint \nabla \cdot \vec{v} \, d\tau$$

Where $d\tau$ is the volume integral.

If $\nabla \cdot \vec{v} = 0$ everywhere, then obviously:

$$\mathcal{V}(t + \Delta t) - \mathcal{V}(t) = 0;$$

or,

$$\mathcal{V}(t) = \text{constant}$$

Implementing this in the étendue conservation, we find that a four-dimensional volume (x, y, p_x, p_y) remains constant, as it evolves according to an optical axis (for exmple direction z). The vector field is $\vec{W} = \left(\dot{x}, \dot{y}, \dot{p}_x, \dot{p}_y\right)$, instead of $\vec{v}$. And $\nabla \cdot \vec{W} = 0$ just as $\nabla \cdot \vec{v} = 0$. This not only implies that the étendue is conserved but also it results in there being no source or sink of field $\vec{W}$ as the light propagates.

2.9 THE GENERALIZED ÉTENDUE OR LAGRANGE INVARIANT AND THE PHASE SPACE CONCEPT

We next have to introduce a concept that is essential to the development of the principles of nonimaging concentrators. We recall that in Section 2.3 we noted that there is a quantity $a^2\theta^2$ that is a measure of the power accepted by the system, where a is the radius of the entrance aperture and θ is the semiangle of the beams accepted. We found that in paraxial approximation for an axisymmetric system this is invariant through the optical system. Actually, we considered only the regions near the entrance and exit apertures, but it is shown in specialized texts on optics that the same quantity can be written down for any region inside a complex optical system. There is one slight complication: if we are considering a region of refractive index different from unity—say, the inside of a lens or prism—the invariant is written as $n^2a^2\theta^2$. The reason for this can be seen from Figure 2.14, which shows a beam at the extreme angle θ entering a plane-parallel plate of glass of refractive index n. Inside the glass the angle is $\theta' = \theta/n$, by the law of refraction,* so that the invariant in this region is

$$\text{étendue} = n^2a^2\theta^2 \tag{2.12}$$

We might try to use the étendue to obtain an upper limit for the concentration ratio of a system as follows. We suppose we have an axisymmetric optical system of any number of components—that is, not necessarily the simple system sketched in Figure 2.7. The system will have an entrance aperture of radius a, which may be the rim of the front lens or, as in Figure 2.16, possibly some limiting aperture inside the system. An incoming parallel beam may emerge parallel, as indicated in the figure, or not, and this will not affect the result. But to simplify the argument it is easier to imagine a parallel beam emerging from an aperture of radius a'. The concentration ratio is by definition $\left(\dfrac{a}{a'}\right)^2$, and if we use the étendue invariant and assume that the initial and final media are both air or vacuum—refractive index unity—the concentration ratio becomes $\left(\dfrac{\theta}{\theta'}\right)^2$. Since from obvious geometrical considerations θ' cannot exceed $\pi/2$, this suggests $\left(\dfrac{2\pi}{\theta}\right)^2$ as a theoretical upper limit to the concentration.

---

* The paraxial approximation is implied so that $\sin\theta \sim \theta$

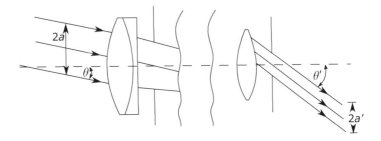

FIGURE 2.16   The étendue for a multielement optical system with an internal aperture stop.

Unfortunately, this argument is invalid because the étendue as we have defined it is essentially a paraxial quantity. Thus, it is not necessarily an invariant for angles as large as $\pi/2$. In fact, the effect of aberrations in the optical system is to ensure that the paraxial étendue is not an invariant outside the paraxial region, so we have not found the correct upper limit to the concentration.

There is, as it turns out, a suitable generalization of the étendue to rays at finite angles to the axis, and we will now explain this. The concept has been known for some time, but it has not been used to any extent in classical optical design, so it is not described in many texts. It applies to optical systems of any or no symmetry and of any structure—refracting, reflecting, or with continuously varying refractive index.

Let the system be bounded by homogeneous media of refractive indices n and n' as in Figure 2.17, and suppose we have a ray traced exactly between the points P and P' in the respective input and output media. We wish to consider the effect of small displacements of P and of small changes in direction of the ray segment through P on the emergent ray so that these changes define a beam of rays of a certain cross section and angular extent. In order to do this we set up a Cartesian coordinate system O_{xyz} in the input medium and another, $O'_{x'y'z'}$, in the output medium. The positions of the origins of these coordinate systems and the directions of their axes are quite arbitrary with respect to each other, to the directions of the ray

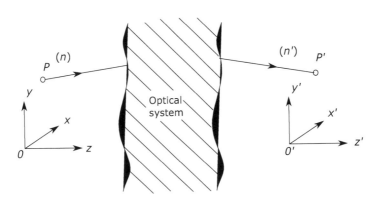

FIGURE 2.17   The generalized étendue.

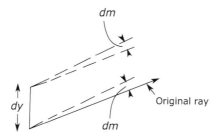

FIGURE 2.18   The generalized étendue in the y section.

segments, and, of course, to the optical system. We specify the input ray segment by the coordinates of $P(x, y, z)$, and by the direction cosines of the ray (L, M, N). The output segment is similarly specified. We can now represent small displacements of P by increments dx and dy in its x and y coordinates, and we can represent small changes in the direction of the ray by increments dL and dM in the direction cosines for the x and y axes. Thus, we have generated a beam of area $dxdy$ and an angular extent defined by $dLdM$. This is indicated in Figure 2.18 for the y section.* Corresponding increments dx', dy', dL', and dM' will occur in the output ray position and direction.

Then the invariant quantity turns out to be $n^2\, dx\, dy\, dL\, dM$—that is, we have

$$n^2\, dx\, dy\, dL\, dM = n^2 dx'\, dy'\, dL'\, dM' \qquad (2.13)$$

The proof of this theorem depends on other concepts in geometrical optics that we do not need in this book. We have given proof using the analogy of statistical mechanics in the previous two sections. Similar proofs can be found in Appendix A. The physical meaning of Equation (2.13) is that it gives the changes in the rays of a beam of a certain size and angular extent as it passes through the system. If there are apertures in the input medium that produce this limited étendue, and if there are no apertures elsewhere to cut off the beam, then the accepted light power emerges in the output medium so that the étendue as defined is a correct measure of the power transmitted along the beam. It may seem at first remarkable that the choice of origin and direction of the coordinate systems is quite arbitrary.

However, it is not very difficult to show that the generalized étendue or Lagrange invariant as calculated in one medium is independent of coordinate translations and rotations. This, of course, must be so if it is to be a meaningful physical quantity.

The generalized étendue is sometimes written in terms of the optical direction cosines $p = nL$, $q = nM$, when it takes the form

$$dx\, dy\, dp\, dq \qquad (2.14)$$

---

* It is necessary to note that the increments dL and dM are in direction cosines, not angles. Thus, in Figure 2.18 the notation on the figure should be taken to mean not that dM is the angle indicated, but merely that it is a measure of this angle.

An étendue value is associated to any four-parameter bundle of rays. Each combination of the four parameters defines one single ray. In the example of Figure 2.17, the four parameters are x, y, L, M (or x', y', L', M'), but there are many other possible sets of four parameters describing the same bundle. For the cases in which the rays are not described at a $z=$ constant (or $z'=$ constant planes), then the following generalized expression can be used to calculate the differential of étendue of the bundle dE:

$$dE = dx\,dy\,dp\,dq + dy\,dz\,dq\,dr + dz\,dx\,dr\,dp \qquad (2.15)$$

The total étendue is obtained by integration of all the rays of the bundles. In what follows we will assume that the bundle can be described at a $z=$ constant plane.

In 2D geometry, when we only consider the rays contained in a plane, we can also define an étendue for any two-parameter bundle of rays. If the plane in which all the rays are contained is an $x=$ constant plane, then the differential of étendue can be written as $dE=n\,dy\,dM$. As in the 3D case, the étendue is an invariant of the bundle, and the same result is obtained no matter where it is calculated. For instance, it can be calculated at $z'=$ constant, and the result should be the same: $n'\,dy'\,dM'=n\,dy\,dM$, or, in terms of the optical direction cosines, $dy'\,dq'=dy\,dq$.

We can now use the étendue invariant to calculate the theoretical maximum concentration ratios of concentrators. Consider first a 2D design, as in Figure 2.19.

We have for any ray bundle that transverses the system

$$n'dy'dM' = n\,dy\,dM \qquad (2.16)$$

and integrating over y and M we obtain

$$4na\sin\theta = 4n'a'\sin\theta' \qquad (2.17)$$

so that the concentration ratio is

$$\frac{a}{a'} = \frac{n'\sin\theta'}{n\sin\theta} \qquad (2.18)$$

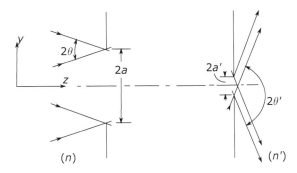

FIGURE 2.19 The theoretical maximum concentration ratio for a 2D optical system.

In this result a' is a dimension of the exit aperture large enough to permit any ray that reaches it to pass, and θ is the largest angle of all the emergent rays. Clearly θ' cannot exceed $\pi/2$, so the theoretical maximum concentration ratio is

$$C_{max} = \frac{a}{a'} = \frac{n'}{n \sin \theta} \qquad (2.19)$$

Similarly, for the 3D case we can show that for an axisymmetric concentrator the theoretical maximum is

$$C_{max} = \left(\frac{a}{a'} \right)^2 = \left(\frac{n'}{n \sin \theta} \right)^2 \qquad (2.20)$$

where again θ is the input semiangle.

The results in Equations (2.19) and (2.20) are maximum values, which may or may not be attained. We find in practice that if the exit aperture has the diameter given by Equation (2.20), some of the rays within the incident collecting angle and aperture do not pass it. We sometimes also find in a number of the systems to be described that some of the incident rays are actually turned back by internal reflections and never reach the exit aperture. In addition, there are losses due to absorption, imperfect reflectivity, and so forth, but these do not represent fundamental limitations.

Thus, Equations (2.19) and (2.20) give theoretical upper bounds on performance of concentrators.

Our results so far apply to linear concentrators, Equation (2.19), with rectangular entrance and exit apertures and to rotational concentrators with circular entrance and exit apertures, Equation (2.20). We ought, for completeness, to discuss briefly what happens if the entrance aperture is not circular but the concentrator itself still has an axis of symmetry. The difficulty with this case is that it depends on the details of the internal optics of the concentrator. It may happen that the internal optical system forms an image of the entrance aperture on the exit aperture—in which case it would be correct to make them similar in shape. For an entry aperture of arbitrary shape but uniform entry angle θ_i all that can be said in general is that for an ideal concentrator the area of the exit aperture must equal that of the entry aperture multiplied by $\sin^2 \theta_i$. We will see in later chapters that such concentrators can be designed.

2.10   THE SKEW INVARIANT

There is an invariant associated with the path of a skew ray through an axisymmetric optical system. Let S be the shortest distance between the ray and the axis—that is, the length of the common perpendicular—and let g be the angle between the ray and the axis. Then the quantity

$$h = nS \sin(\gamma) \qquad (2.21)$$

is an invariant through the whole system. If the medium has a continuously varying refractive index, the invariant for a ray at any coordinate z_1 along the axis is obtained by treating the tangent of the ray at the z value as the ray and using the refractive index value at the point where the ray cuts the transverse plane z_1. If we use the dynamical analogy described in Appendix A, then h corresponds to the angular momentum of a particle following the ray path, and the skew invariant theorem corresponds to conservation of angular momentum. In terms of Hamilton's equations, the skew invariant is just a first integral that derives from the symmetry condition.

2.11 DIFFERENT VERSIONS OF THE CONCENTRATION RATIO

We now have some different definitions of concentration ratio. It is desirable to clarify them by using different names. First, in Section 2.7 we established upper limits for the concentration ratio in 2D and 3D systems, given respectively by Equations (2.14) and (2.15). These upper limits depend only on the input angle and the input and output refractive indices. Clearly we can call either expression the theoretical maximum concentration ratio.

Second, an actual system will have entry and exit apertures of dimensions $2a$ and $2a'$. These can be width or diameter for linear or rotational systems, respectively. The exit aperture may or may not transmit all rays that reach it, but in any case the ratios (a/a') or $(a/a')^2$ define a geometrical concentration ratio.

Third, given an actual system, we can trace rays through it and determine the proportion of incident rays within the collecting angle that emerge from the exit aperture. This process will yield an optical concentration ratio.

Finally, we could make allowances for attenuation in the concentrator by reflection losses, scattering, manufacturing errors, and absorption in calculating the optical concentration ratio. We could call the result the optical concentration ratio with allowance for losses. The optical concentration ratio will always be less than or equal to the theoretical maximum concentration ratio. The geometrical concentration ratio can, of course, have any value.

REFERENCES

Luneburg, R. K. (1964). *Mathematical Theory of Optics*. University of California Press, Berkeley, CA.
Welford, W. T. (1974). *Aberrations of the Symmetrical Optical System*. Academic Press, New York.
Welford, W. T. (1986). *Aberrations of Optical Systems*. Hilger, Bristol, England.

3 Some Designs of Image-Forming Concentrators

3.1 INTRODUCTION

In this chapter we examine image-forming concentrators of conventional form—paraboloidal mirrors, lenses of short focal length, and so forth—and estimate their performance. Then we shall show how the departure from ideal performance suggests a principle for the design of nonimaging concentrators—the "edge-ray principle," as we shall call it.

3.2 SOME GENERAL PROPERTIES OF IDEAL IMAGE-FORMING CONCENTRATORS

In order to fix our ideas, we use the solar energy application to describe the mode of action of our systems. The simplest hypothetical image-forming concentrator would then function as in Figure 3.1. The rays are coded to indicate that rays from one direction from the sun are brought to a focus at one point in the exit aperture—that is, the concentrator images the sun (or other source) at the exit aperture. Notice that the rays entering the concentrator at a certain angle positive θ, marked with the single arrow, will end up at the bottom imaging point at the exit. In other words, the rays from the upper edge of the infinitely far away object are arriving the bottom pixel of the image.

If the exit medium is air, then the exit angle θ must be $\pi/2$ for maximum concentration. Such a concentrator may in practice be constructed with glass or some other medium of refractive index greater than unity forming the exit surface, as in Figure 3.2. Also, the angle θ in the glass would have to be such that $\sin\theta = 1/n$ so that the emergent rays just fill the required $\pi/2$ angle. For typical materials the angle θ would be about 40°.

Figure 3.2 brings out an important point about the objects of such a concentrator.

We have labeled the central or principal ray of the two extreme angle beams a and b, respectively, and at the exit end these rays have been drawn normal to the exit face. This would be essential if the concentrator were to be used with air as the final medium, since, if rays a and b were not normal to the exit face, some of the extreme angle rays would be totally internally reflected (see Section 2.2), and thus the concentration ratio would be reduced. In fact, the condition that the exit principal rays should be normal to what, in ordinary lens design, is termed the image plane is not usually fulfilled. Such an optical system, called telecentric, needs to be specially

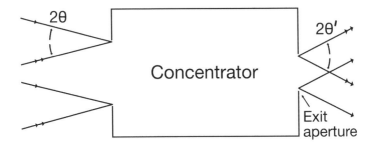

FIGURE 3.1   An image-forming concentrator. An image of the source at infinity is formed at the exit aperture of the concentrator.

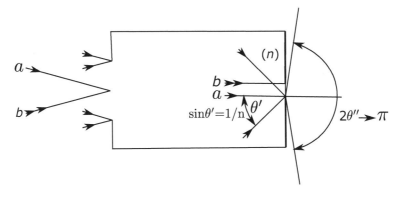

FIGURE 3.2   In an image-forming concentrator of maximum theoretical concentration ratio the final medium in the concentrator would have to have a refractive index n greater than unity. The angle θ in this medium would be arcsin $(1/n)$, giving an angle $\pi/2$ in the air outside.

designed, and the requirement imposes constraints that would certainly worsen the attainable performance of a concentrator. We shall therefore assume that when a concentrator ends in glass of index n, the absorber or other means of utilizing the light energy is placed in optical contact with the glass in such a way as to avoid potential losses through total internal reflection. An alternative configuration for an image-forming concentrator would be as in Figure 3.3. The concentrator collects

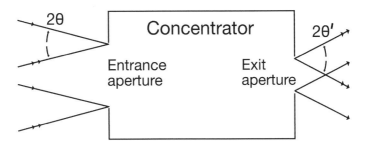

FIGURE 3.3   An alternative configuration of an image-forming concentrator. The rays collected from an angle θ form an image of the entrance aperture at the exit aperture.

rays over θ_{max} as before, but the internal optics form an image of the entrance aperture at the exit aperture, as indicated by the arrow coding of the rays. This would be in optics terminology a telescopic or afocal system. Such systems simply move the object from infinitely far away to position of finite distance. Naturally, the same considerations about using glass or a similar material as the final medium holds as for the system of Figures 3.1 and 3.2, and there is no difference between the systems as far as external behavior is concerned, with regards to the possible exit angles and without any loss introduced by the total internal reflections.

If the concentrator terminates in a medium of refractive index n, we can gain in maximum concentration ratio by a factor n or n^2, depending on whether it is a 2D or 3D system, as can be seen from Equations (2.13) and (2.14). This corresponds to having an extreme angle $\theta' = \pi/2$ in this medium. We then have to reinstate the requirement that the principal rays be normal to the exit aperture, and we also have to ensure that the absorber can utilize rays of such extreme angles.

In practice there are problems in using extreme collection angles approaching $\theta = \pi/2$ whether in air or a higher-index medium. There has to be very good matching at the interface between glass and absorber to avoid large reflection losses of grazing-incidence rays, and irregularities of the interface can cause losses through shadowing. Therefore, we may well be content with values of θ' of, say, 60°. This represents only a small decrease from the theoretical maximum concentration, as can be seen from Equations (2.14) and (2.15).

Thus, in speaking of ideal concentrators we can also regard as ideal a system that brings all incident rays within θ_{max} out within θ'_{max} and inside an exit aperture a' given by Equation (2.12)—that is, $a' = na \sin\theta_{max} / n' \sin\theta'_{max}$. Such a concentrator would be ideal, but it will not have the theoretical maximum concentration.

The concentrators sketched in Figures 3.1 and 3.2 clearly must contain something like a photographic objective with very large aperture (small f-number), or perhaps a high-power microscope objective used in reverse. The speed of a photographic objective is indicated by its f-number or aperture ratio. Thus, an $f/4$ objective has a focal length four times the diameter of its entrance aperture. This description is not suitable for imaging systems in which the rays form large angles approaching $\pi/2$ with the optical axis for a variety of reasons. It is found that in discussing the resolving power of such systems the most useful measure of performance is the numerical aperture or NA, a concept introduced by Ernst Abbe in connection with the resolving power of microscopes. Figure 3.4 shows an optical system with an entrance aperture of diameter $2a$. It forms an image of the axial object point at infinity and the semi-angle of the cone of extreme rays is a θ_{max}. Then the numerical aperture is defined by

$$NA = n' \sin\alpha'_{max} \tag{3.1}$$

where n' is the refractive index of the medium in the image space. We assume that all the rays from the axial object point focus sharply at the image point—that is, there is (to use the terminology of Section 2.4) no spherical aberration. Then Abbe showed that off-axis object points will also be sharply imaged if the condition

$$h = n' \sin\alpha' \times \text{const.} \tag{3.2}$$

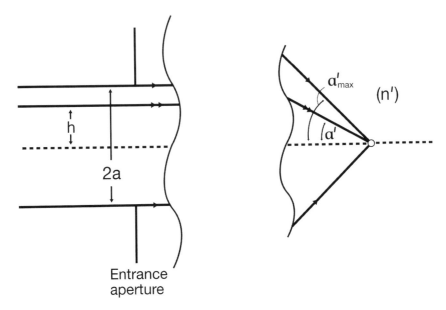

FIGURE 3.4   The definition of the numerical aperture of an image-forming system. The NA is $n' \sin \alpha'$.

is fulfilled for all the axial rays. In this equation h is the distance from the axis of the incoming ray, and α' is the angle at which that ray meets the axis in the final medium. Equation (3.2) is a form of the celebrated Abbe sine condition for good image formation. It does not ensure perfect image formation for all off-axis object points, but it ensures that aberrations that grow linearly with the off-axis angle are zero. These aberrations are various kinds of coma. The condition of freedom from spherical aberration and coma is called aplanatism.

Clearly, a necessary condition for our image-forming concentrator to have the theoretical maximum concentration—or even for it to be ideal as an image-forming system (but without theoretical maximum concentration)—is that the image formation should be aplanatic. This is not, unfortunately, a sufficient condition.

The constant in Equation (3.2) has the significance of a focal length. The definition of focal length for optical systems with media of different refractive indices in the object and image spaces is more complicated than for the thin lenses discussed in Chapter 2. In fact, it is necessary to define two focal lengths, one for the input space and one for the output space, where their magnitudes are in the ratio of the refractive indices of the two media. In Equation (3.2) it turns out that the constant is the input side focal length, which we shall denote by f.

From Equation (3.2) we have for the input semiaperture

$$a = f \cdot NA \tag{3.3}$$

and also, from Equation (2.118),

$$a' = a \sin \theta_{max} / NA \tag{3.4}$$

By substituting from Equation (3.3) into Equation (3.4) we have

$$a' = f \sin \theta_{\max} \qquad (3.5)$$

where $\theta_{\max}$ is the input semiangle. To see the significance of this result we recall that we showed that in an aplanatic system the focal length is a constant, independent of the distance h of the ray from the axis used to define it. Here we are using the generalized sense of "focal length" meaning the constant in Equation (3.2), and aplanatism thus means that rays through all parts of the aperture of the system form images with the same magnification. Thus, Equation (3.5) tells us that in an imaging concentrator with maximum theoretical concentration the diameter of the exit aperture is proportional to the sine of the input angle. This is true even if the concentrator has a numerical exit aperture less than the theoretical maximum, n', provided it is ideal in the sense just defined.

From the point of view of conventional lens optics, the result of Equation (3.3) is well known. It is simply another way of saying that the aplanatic lens with the largest aperture and with air as the exit medium is $f/0.5$, since Equation (3.5) tells us that $a=f$. The importance of Equation (3.5) is that it tells us something about one of the shape-imaging aberrations required of the system—namely distortion. A distortion-free lens imaging onto a flat field must obviously have an image height proportional to $\tan \theta$, so our concentrator lens system is required to have what is usually called barrel distortion. This is illustrated in Figure 3.5. Our picture of an imaging concentrator is gradually taking shape, and we can begin to see that certain requirements of conventional imaging can be relaxed.

Thus, if we can get a sharp image at the edge of the exit aperture and if the diameter of the exit aperture fulfils the requirement of Equations (3.3)–(3.5), we do not need perfect image formation for object points at angles smaller than $\theta_{\max}$. For example, the image field perhaps could be curved, provided we take the exit aperture in the plane of the circle of image points for the direction $\theta_{\max}$, as in Figure 3.6. Also, the inner parts of the field could have point-imaging aberrations, provided these were not so large as to spill rays outside the circle of radius a'. Thus, we see that an image-forming concentrator need not, in principle, be as difficult to design as an imaging lens, since the aberrations need to be corrected only at the edge of the field. In practice this relaxation may not be very helpful because the outer part of the field is the most difficult to correct. However, this leads us to a valuable principle for nonimaging concentrators. Not only is it unnecessary to have good aberration correction except at the exit rim, but we do not even need point imaging at the rim itself. It is only necessary that rays entering at the extreme angle $\theta_{\max}$ leave from some point at the rim and that the aberrations inside do not push rays outside the rim of the exit aperture. We shall return to this edge-ray principle later in connection with nonimaging concentrators. The above arguments need only a little modification to apply to the alternative configuration of an imaging concentrator in Figure 3.3, in which the entrance aperture is imaged at the exit aperture. Referring to Figure 3.7, we can imagine that the optical components of the concentrator are forming an image at the exit aperture of an object at a considerable distance, rather than at infinity, and that this object is the entrance aperture. Alternatively, we can imagine that part of the

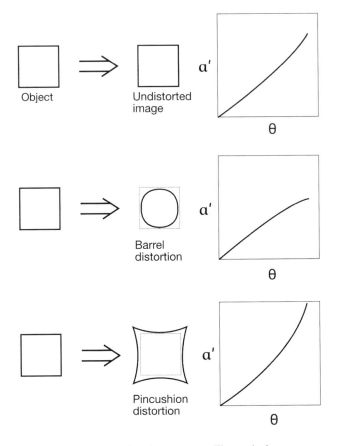

FIGURE 3.5 Distortion in image-forming systems. The optical systems are assumed to have symmetry about an axis of rotation.

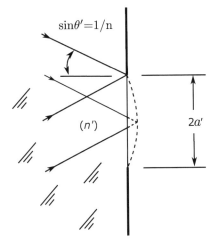

FIGURE 3.6 A curved image field with a plane exit aperture.

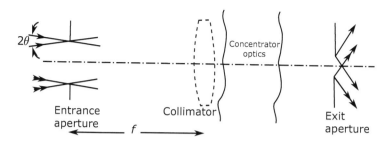

FIGURE 3.7 An afocal concentrator shown as two image-forming systems.

concentrator is a collimating lens of focal length f, shown in broken line in the figure, and that this projects the entrance aperture to infinity with an angle subtending $2a/f$. The same considerations as before then apply to the aberration corrections.

3.3 CAN AN IDEAL IMAGE-FORMING CONCENTRATOR BE DESIGNED?

In Section 3.2 we outlined some requirements for ideal image-forming concentrators, and we now have to ask whether they can be designed to fulfill these requirements with useful collection angles.

High-aperture camera lenses are made at about $f/1.0$, but these are complex structures with many components. Figure 3.8 shows a typical example with a focal length of 50 mm. Such a system is by no means aberration-free, and the cost of scaling it up to a size useful for solar work would be prohibitive. Anyway, its numerical aperture is still only about 0.5. The only systems with numerical apertures approaching the theoretical limit are microscope objectives. Figure 3.9 shows one of the simplest designs of microscope objective of numerical aperture about 1.35, drawn in reverse and with one conjugate at infinity. The image or exit space has a refractive index of 1.52, since it is an oil immersion objective. Such systems have good aberration correction only to about 3° from the axis. Beyond this the aberrations increase rapidly, and also there is less light transmission because of vignetting.* The collecting aperture would be about 4 mm in diameter. Again, it would be impracticable to scale up such a system to useful dimensions.

Thus, a quick glance at the state of the art in conventional lens design suggests that imaging concentrators in the form of lens systems will not be very efficient on a practical scale. Nevertheless, it is interesting to see what might be done with the classical imaging design techniques if practical limitations are ignored.

Roughly, the position seems to be that we cannot design an ideal concentrator—one with the theoretical maximum collection efficiency—using a finite number of lens elements. But, by increasing the number of elements sufficiently or by postulating sufficiently extreme optical properties, we can approach indefinitely close to the ideal. Exceptions to the preceding proposed rule occur in optical systems with

---

* Vignetting is caused by rims of components at either end of a long system shearing against each other as the system is turned off-axis.

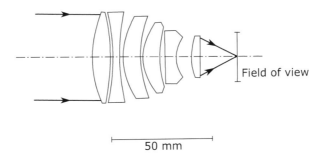

FIGURE 3.8 A high-aperture camera objective. The drawing is to scale for a 50-mm focal length. The emerging cone of rays has a semiangle of 26° at the center of the field of view.

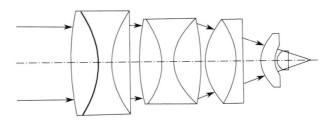

FIGURE 3.9 An oil-immersion microscope objective of high numerical aperture. Such systems can have a convergence angle of up to 60° with an aberration-free field of about ±3°. However, they can only be designed aberration-free for focal lengths up to 2 mm—that is, an actual field diameter of about 200 mm.

spherical symmetry. It has been known since the time of Huygens that a spherical lens element images a concentric surface, as shown in Figure 3.10.

The two conjugate surfaces have radii r/n and nr, respectively. The configuration is used in microscope objectives having a high numerical aperture, as in Figure 3.9. Unfortunately, one of the conjugates must always be virtual (the object conjugate as the figure is drawn), so the system alone would not be very practical as a concentrator. It seems to be true, although this has not been proven, that no combination of a finite number of concentric components can form an aberration-free real image of a real object. However, as we shall see, this can be done with media of continuously varying refractive index. The system of Figure 3.10 would clearly be useful as the last stage of an imaging concentrator. It can easily be shown that the convergence angles are related by the equation

$$\sin\alpha' = n\sin\alpha \qquad (3.6)$$

Also, if there is a plane surface terminating in air, the final emergent angle α'' is given by

$$\sin\alpha'' = n^2\sin\alpha \qquad (3.7)$$

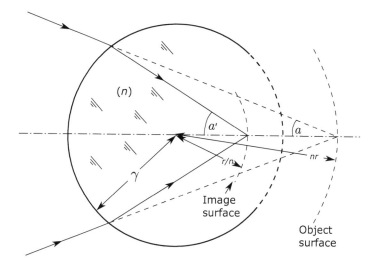

FIGURE 3.10 The aplanatic surfaces of a spherical refracting surface.

Thus, the system could be used in conjunction with another system of relatively low numerical aperture, as in Figure 3.11, to form a fairly well-corrected concentrator. This is, of course, merely a reinvention of the microscope objective of Figure 3.9, and the postulated additional system still needs to operate at about $f/1$ if ordinary materials are used. If we assume some extreme material qualities—say, a refractive index of 4 with adequate antireflection coating for the aplanatic component—then the auxiliary system only needs to be $f/8$ to give $\pi/2$ emergent angles in air, as in Figure 3.12.

This is not a difficult requirement. In fact, it is probably true that if we ignore chromatic aberration—that is, if we assume our postulated material of index 4 has no dispersion—this system could be designed in moderate sizes for which the aberrations are indefinitely small over a reasonable acceptance angle.

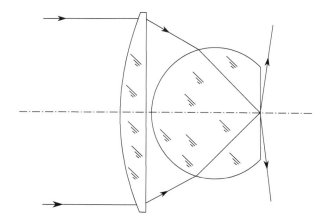

FIGURE 3.11 An image-forming concentrator with an aplantic component.

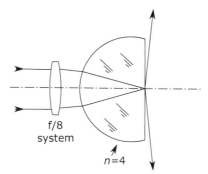

FIGURE 3.12 Use of an aplanatic component of high refractive index to produce a well-corrected optical system.

Ultimately the ray aberrations of optical systems become negligible because the performance, as an imaging system, is limited by diffraction effects. Thus, if an imaging optical system of a certain numerical aperture is used to form an image of a point source and if there are precisely no ray aberrations, the image of the point will not be indefinitely small. It is shown in books on physical optics[*] that the point image will be a blurred diffraction pattern in which most of the light flux falls inside a circle of radius

$$0.61\lambda \,/\, NA \tag{3.8}$$

where λ is the wavelength of the light. Notice the first zero of the diffraction pattern (Airy pattern) happens at $1.22\ \lambda/D$, where D is the size of the aperture. This provides us with a tolerance level for ray aberrations below which we can say the aberrations are negligible. This is sometimes expressed in the form that all of the points in the spot diagram for the aberrations (see Figure 2.10) must fall within a circle of radius given by Equation (3.8).

Another way of setting a tolerance is to say that the wave-front shape as determined by the methods outlined in Section 2.6 would not depart from the ideal spherical shape by more than a specified amount, usually $\lambda/4$. Other tolerance systems are also described by Born and Wolf (1975).

Thus, in some way we could arrive at tolerances for the geometrical aberrations such that an imaging concentrator with aberrations inside these tolerances would have ideal performance. Any lens system made with available materials would be impractically complicated and costly if scaled up to a size suitable for solar concentration. But hypothetical materials could be used to bring the aberrations within the diffraction limit with a simple design such as in Figure 3.12.

At this point it might be interesting to consider a different example of a lens system as a concentrator. It is well known that an ellipsoidal solid lens will focus parallel light without spherical aberration, as in Figure 3.13, provided the eccentricity is equal to $1/n$. In order to use this as a concentrator we put the entry aperture in front of

---

[*] See, for example, Born and Wolf, 1975.

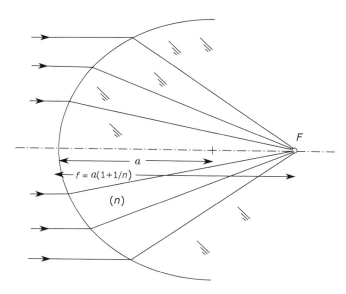

FIGURE 3.13 A portion of an ellipsoid of revolution as a single refracting surface free from spherical aberration. The generating ellipse has eccentricity $1/n$ and semi-major axis a. The figure is drawn to scale with $n = 1.81$.

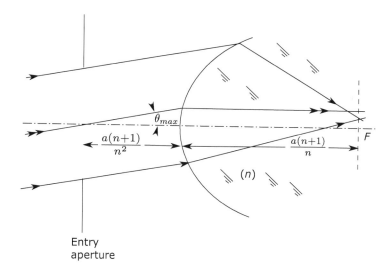

FIGURE 3.14 An ellipsoid of revolution as a concentrator. The entry aperture is set at the first focus of the system.

the ellipsoid, as in Figure 3.14, so that the principal ray emerges parallel to the axis. We find that the ellipsoid has very strong coma of such a sign that all other rays meet the image plane nearer the axis than the principal ray.

Thus, all rays strike within this circle. However, this is not an ideal concentrator, since, as can be seen from the diagram, the radius of this circle is proportional to

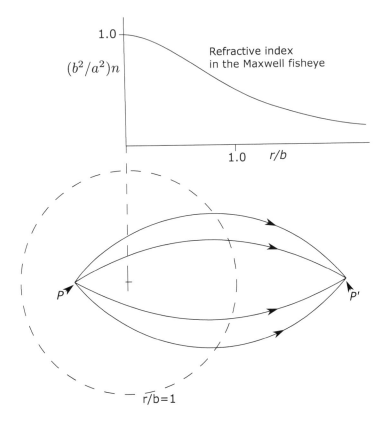

FIGURE 3.15   Rays in the Maxwell fisheye. The rays are arcs of circles.

$\tan(\theta_{max})$ whereas, according to Equation (3.5), all rays should strike within a circle of radius proportional to $\sin(\theta_{max})$.[*]

3.4   MEDIA WITH CONTINUOUSLY VARYING REFRACTIVE INDICES

We stated in Section 3.3 that it is thought to be impossible to design an ideal imaging concentrator with a finite number of reflecting or refracting surfaces, even with spherical symmetry, although the example of Figure 3.10 shows that perfect imagery is possible if one conjugate surface is virtual. It has long been known that if we admit continuously varying refractive index, then perfect imagery between surfaces in a spherically symmetric geometry is possible. James Clerk Maxwell (1854) showed that if a medium had the refractive index distribution (Figure 3.15)

$$n = \frac{a^2}{b^2 + r^2} \tag{3.9}$$

---

[*] In fact, on account of the curvature of the ellipse, the radius is even slightly greater than a value proportional to $\tan(\theta_{max})$.

where a and b are constants and r is a radial coordinate, then any point would be perfectly imaged at another point on the opposite side of the origin. If $a=b=1$, the distances of conjugate points from the origin are related by

$$rr' = 1 \tag{3.10}$$

immersed in the medium. Luneburg (1964) gave several more examples of media of spherical symmetry with ideal imaging properties. In particular, he found an example, now known in the literature as the Luneburg lens, where the index distribution extends over a finite radius only and where the object conjugate is at infinity. The index distribution is

$$n(r) = \begin{cases} \sqrt{2 - \dfrac{r^2}{a^2}}, & r < 1 \\ 1, & r \geq 1 \end{cases} \tag{3.11}$$

This distribution forms a perfect point image with numerical aperture unity, as in Figure 3.16, and on account of the spherical symmetry, it can be shown to form an ideal concentrator of maximum theoretical concentration.

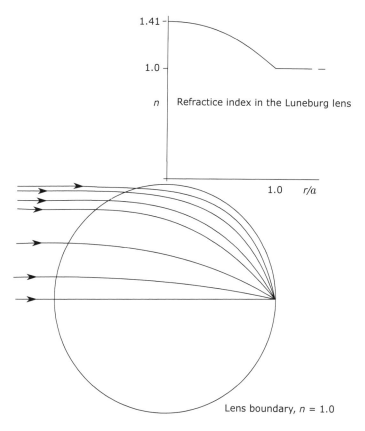

FIGURE 3.16   The Luneburg lens.

Appendix B presents a more detailed treatment that shows how the ray paths are calculated. You can see that the Luneburg lens satisfied Abbe's sine condition (Equation [3.2]). Also, it follows from the spherical symmetry that perfect point images are formed from parallel rays coming in all directions. It is then possible to consider the Luneburg lens as having the theoretical maximum concentration ratio for any desired collection angle θ_{max} up to $\pi/4$ but collecting from a concave spherical source at infinity onto a concave spherical absorber attached to the lens. Apart from the practical problem of making the lens this is rather an artificial configuration, since up until now we have been considering plane entry and exit apertures. Yet, the Luneburg lens would have an exit aperture in the form of a spherical cap and an entrance aperture that changes in shape with the angle of the rays. Nevertheless, we show in Appendix B that with reasonable and consistent interpretations of "entrance aperture" and "exit aperture" the Luneburg lens has an optical concentration ratio equal to the theoretical maximum.

3.5 ANOTHER SYSTEM OF SPHERICAL SYMMETRY

The discussion in the last section, in which it was suggested that the ideas of concentration could be extended to nonplane absorbers, suggests a way in which the aplanatic imaging system of Figure 3.10 could be used by itself as a concentrator, as in Figure 3.17. The surface of radius a/n forms the spherical exit surface, and the internal angle 2α of the cone meeting this face is such that $\sin \alpha = 1/n$. Thus, the emerging rays cover a solid angle 2π, as with the Luneburg lens. The entry aperture is now a virtual aperture on the surface of radius an. The collecting angle θ_{max} is thus given by $\sin \theta_{max} = 1/n^2$. The concentration ratio from air to air is n^4 for a 3D system—that is, it is determined only by the refractive index. Similarly, the collecting angle is fixed.

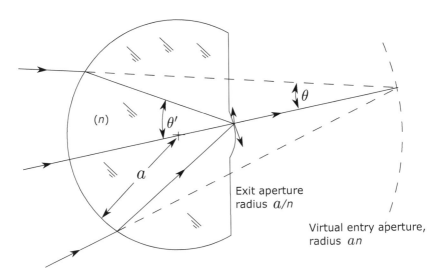

FIGURE 3.17 The aplanatic spherical lens as an ideal concentrator; the diagram is to scale for refractive index $n=2$; the rays shown emerge in air as the extreme rays in a solid angle 2π.

Thus, this is not a very flexible system, apart from the fact that it has a virtual collecting aperture. But it does have the theoretical maximum concentration ratio, and, if we admit such systems, it is another example of an ideal system.

3.6 IMAGE-FORMING MIRROR SYSTEMS

In this section we examine the performance of mirror systems as concentrators. Concave mirrors have, of course, been used for many years as collectors for solar furnaces and the like. Historical material about such systems is given by Krenz (1976). However, little seems to have been published in the way of angle transmission curves for such systems. Consider first a simple paraboloidal mirror, as in Figure 3.18. As is well known, this mirror focuses rays parallel to the axis exactly to a point focus, or, in our terminology, it has no spherical aberration.

However, the off-axis beams are badly aberrated. Thus, in the meridian section (the section of the diagram) it is easily shown by ray-tracing that the edge rays at angle θ meet the focal plane further from the axis than the central ray, so this cannot be an ideal concentrator even for emergent rays at angles much less than $\pi/2$. An elementary geometrical argument (see, e.g., Harper et al., 1976) shows how big the exit aperture must be to collect all the rays in the meridian section.

Referring to Figure 3.19, we draw a circle passing through the ends of the mirror and the absorber (i.e., exit aperture). Then, by a well-known property of the circle, if the absorber subtends an angle $4\theta_{max}$ at the center of the circle, it subtends $2\theta_{max}$ at the ends of the mirror, so the collecting angle is $2\theta_{max}$. The mirror is not specified

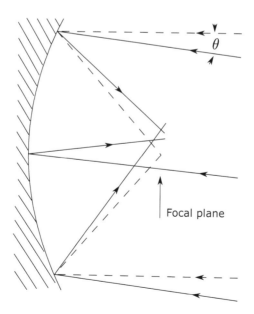

FIGURE 3.18 Coma of a paraboloidal mirror. The rays of an axial beam are shown in broken line. The outer rays from the oblique beams at angle q meet the focal plane further from the axis than the central ray of this beam.

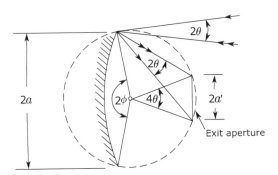

FIGURE 3.19   Collecting all the rays from a concave mirror.

to be of any particular shape except that it must reflect all inner rays to the inside of the exit aperture. Then if the mirror subtends 2ϕ at the center of the circle, we find

$$\frac{a'}{a} = \frac{\sin 2\theta_{max}}{\sin \phi} \tag{3.12}$$

and the minimum value of a' is clearly attained when $\phi = \pi/2$. At this point the optical concentration ratio is, allowing for the obstruction caused by the absorber,

$$\left(\frac{a'}{a}\right)^2 - 1 = \frac{1}{4\sin^2 \theta_{max}} \cdot \frac{\cos^2 2\theta_{max}}{\cos^2 \theta_{max}} \tag{3.13}$$

It can be seen that this is less than 25% of the theoretical maximum concentration ratio and less than 50% of the ideal for the emergent angle used. If, as is usual, the mirror is paraboloidal, the rays used for this calculation are actually the extreme rays—that is, the rays outside of the plane of the diagram all fall within the circle of radius a.

The large loss in concentration at high apertures is basically because the single concave mirror used in this way has large coma—in other words, it does not satisfy Abbe's sine condition Equation (3.2). The large amount of coma introduced into the image spreads the necessary size of the exit aperture and so lowers the concentration below the ideal value.

There are image-forming systems that satisfy the Abbe sine condition and have large relative apertures. The prototype of these is the Schmidt camera, which has an aspheric plate and a spherical concave mirror, as shown in Figure 3.20. The aspheric plate is at the center of curvature of the mirror, and thus the mirror must be larger than the collecting aperture. Such a system would have the ideal concentration ratio for a restricted exit angle apart from the central obstruction, but there would be practical difficulties in achieving the theoretical maximum. In any case a system of this complexity is clearly not to be considered seriously for solar work.

3.7   CONCLUSIONS ON CLASSICAL IMAGE-FORMING CONCENTRATORS

It must be quite clear by now that, whatever the theoretical possibilities, practical concentrators based on classical image-forming designs fall a long way short of the

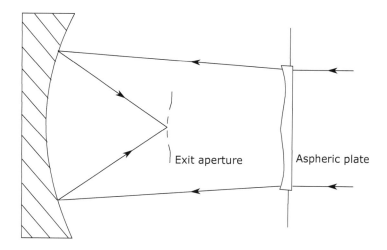

FIGURE 3.20 The Schmidt camera. This optical system has no spherical aberration or coma, so, in principle, it could be a good concentrator for small collecting angles. However, there are serious practical objections, such as cost and the central obstruction of the aperture.

ideal. Our graphs of angular transmission will indicate this for some of the simpler designs. As to theoretical possibilities, it is certainly possible to have an ideal concentrator of theoretical maximum concentration ratio if we use a spherically symmetric geometry, a continuously varying refractive index, and quite unrealistic material properties (i.e., refractive index between 1 and 2 and no dispersion).

This was proved by the example of the Luneburg lens, and Luneburg and others (e.g., Morgan, 1958; Cornbleet, 1976) have shown how designs suitable for perfect imagery for other conjugates can be obtained.

Nevertheless, imaging and nonimaging are not opposite concepts. Only if we restrict ourselves to a finite number of elements can perfect concentrators with plane apertures and axial symmetry not be obtained.

REFERENCES

Born, M., and Wolf, E. (1975). *Principles of Optics*, 5th Ed. Pergamon, Oxford.

Cornbleet, S. (1976). *Microwave Optics*. Academic Press, New York.

Harper, D. A., Hildebrand, R. H., Pernic, R., and Platt, R. (1976). Heat trap: An optimised far infrared field optics system. *Appl. Opt.* **15**, 53–60.

Krenz, J. H. (1976). *Energy Conversion and Utilization*. Allyn & Bacon, Rockleigh, NJ.

Luneburg, R. K. (1964). *Mathematical Theory of Optics*. University of California Press, Berkeley, CA. This material was originally published in 1944 as loose sheets of mimeographed notes and the book is a word-for-word transcription.

Maxwell, J. C. (1958). On the general laws of optical instruments. *Q. J. Pure Appl. Math.* **2**, 233–247.

Morgan, S. P. (1958). General solution of the Luneburg lens problem. *J. Appl. Phys.* **29**, 1358–1368.

4 Nonimaging Optical Systems

4.1 LIMITS TO CONCENTRATION

The relationship between the concentration ratio and the angular field of view is a fundamental one that merits more than one demonstration. We shall give a thermodynamic argument in the context of solar energy concentration. Imagine the sun itself as a spherically symmetric source of radiant energy (Figure 4.1). The flux falls off as the inverse square of the distance R from the center, as follows from the conservation of power through successive spheres of area $4\pi R^2$. Therefore, the flux on the earth's surface, say, is smaller than the solar surface flux by a factor $(r/R)^2$, where r is the radius of the sun and R is the distance from the earth to the sun. By simple geometry, $r/R = \sin^2\theta$ where θ is the angular subtense (half angle) of the sun. If we accept the premise that no terrestrial device can boost the flux above its solar surface value (which would lead to a variety of perpetual motion machines), then the limit to concentration is just $1/\sin^2\theta$.

We call this limit the sine law of concentration. This relationship may seem similar to the well-known Abbe sine condition of optics, but the resemblance is only superficial. The Abbe condition applies to well-corrected optical systems and is the first order in the transverse dimensions of the image. There are no such limitations to the sine law of concentration, which is correct and rigorous for any size receiver. As already shown in Equations (2.14) and (2.15), there is an escape clause to this conclusion when the target is immersed in a medium with index of refraction (n) because then the limit is $n^2/\sin\theta^2$. Of course, the limit we have derived is for concentration in both transverse dimensions, which we will refer to as three-dimensional concentration (or 3D concentration for short). For concentration in one transverse dimension, which we will call two-dimensional (2D) concentration, the limit is clearly $1/\sin\theta$. While the concept of concentration in our demonstration refers to solar flux, implicit in our discussion is good energy throughput. We generally deal with concentrators that throw away as little energy as possible. Good energy conservation is an essential attribute of a useful concentrating optical system (see Figure 4.1).

$$D = 2r\sin\phi$$

$$d = \frac{2r\sin\theta}{\cos\phi}$$

$$\frac{D}{d} = \frac{\sin\phi\cos\phi}{\sin\theta} = \frac{\sin 2\phi}{2\sin\theta}$$

$$C = \left(\frac{D}{d}\right)^2 = \frac{1}{4}\cdot\frac{\sin^2 2\phi}{\sin^2\theta_s} \le \frac{1}{4}\cdot\frac{1}{\sin^2\theta_s} \le \left(\frac{1}{4}\right)C_{max}$$

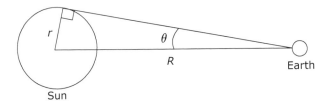

FIGURE 4.1 The flux from a spherically symmetric sun falls off $\left(\dfrac{r}{R}\right)^2 = \left(\dfrac{1}{\sin\theta}\right)^2$.

4.2 IMAGING DEVICES AND THEIR LIMITATIONS

If one were to ask the proverbial man on the street for a suggestion of how one might attain the highest possible level of concentration of, say, solar flux, a plausible response would be to use a good astronomical telescope—perhaps the 200-inch telescope on Mt. Palomar or any preferred telescope. Of course, such an experiment had better remain in the realm of imagination only, since beginner astronomers are would be admonished for pointing their telescopes at the sun at the risk of catastrophic damage to the instrument. But to continue this train of thought, the concentration limit of a telescope is readily shown to be $\dfrac{\sin^2 2\phi}{4\sin^2\theta}$ after an elementary calculation (Figure 4.2). Here we have introduced a new parameter, ϕ, which is the rim angle of the telescope. The best one can do is make the numerator 1 for rim angle $f=45°$, so the best concentration achieved is $\dfrac{1}{4\sin^2\theta}$, which falls short of the fundamental limit by a factor of 4! Now factors of 4 are significant in technology (and

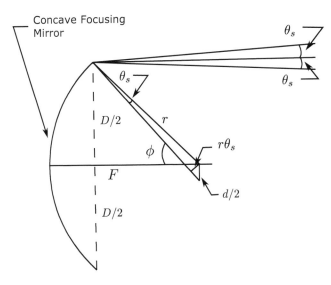

FIGURE 4.2 Image-forming solar concentrator for planar absorber.

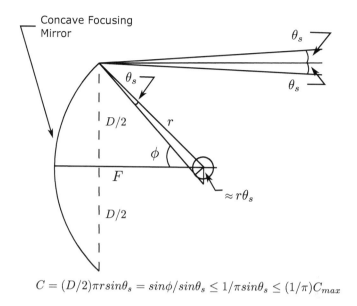

$$C = (D/2)\pi r sin\theta_s = sin\phi/sin\theta_s \le 1/\pi sin\theta_s \le (1/\pi)C_{max}$$

FIGURE 4.3   Image-forming solar concentrator for tubular absorber.

many other forms of human endeavor). It was the desire to bridge the gap between the levels of concentration achieved by common imaging devices and the sine law of concentration limit that motivated the invention of nonimaging optics. Entirely similar considerations can be applied to 2D or trough concentrators. A straightforward generalization to a strip absorber rather than a disk absorber gives a limit for, say, a parabolic trough of $\dfrac{sin 2\phi}{2 sin \theta}$, with a consequent upper limit of $\dfrac{1}{2 sin \theta}$ for rim angle $\phi = 45°$. This would be a useful configuration for a photovoltaic concentrator, with the strip consisting of solar cells. However, a more useful geometry for a parabolic trough thermal concentrator is a tubular receiver (Figure 4.3). In that case, the concentration relation becomes $\dfrac{sin \phi}{\pi sin \theta}$, which attains its maximum value of $\dfrac{1}{\pi sin \theta}$ at a 90° rim angle.

This helps to explain why parabolic solar troughs tend to have large rim angles.

In either case, we fall significantly short of the sine law of concentration limit, this time by a factor of 2 or π, depending on the configuration. As we shall see later, this factor of 2 or π makes the difference between the possibility and impossibility of fixed solar concentrators. The possibility of fixed concentrators opens up a broad vista of solar energy applications.

4.3   NONIMAGING CONCENTRATORS

These simple examples of imaging systems and their attendant shortfalls in concentrating performance (we could have examined lenses and reached similar conclusions) suggest that the requirement that an optical concentrator form an image is unduly restrictive. After all, we are after transport of radiant energy. Imaging the sun may be useful in solar astronomy or in the study of sun spots, but it has no obvious advantage in solar energy conversion systems. Thus, even taking a more empirical optimization approach, it is plausible that relaxing the imaging requirement has the potential of improving concentrating performance; in other words, one would expect to be able to trade off one against the other. Approaching the subject this way does lead to incremental improvements over various classical imaging designs such as parabolic reflectors. Our approach in this book is to show methods that actually attain, or closely approach, the theoretical sine law limit to concentration while maintaining high throughput, methods that bear little resemblance to classical imaging approaches. An analogy with fluid dynamics may be useful to bring this point home. In fluid dynamics, as in optics, a useful representation is in "phase space." Phase space has twice the dimensions of ordinary space and consists of both the positions and momenta of elements of the fluid. In optics, the momenta are the directions of light rays multiplied by the index of refraction of the medium. In optics, as in fluid dynamics, the volume in this phase space is conserved, a sort of incompressible fluid flowing in this space of twice the number of physical dimensions. Now consider an imaging problem taking the simplest example of points on a line. An imaging system is required to map those points on another line, called the image, without scrambling the points (Figure 4.4a)—that is, to focus the rays issuing from every object point into their corresponding image points. All the rays issuing from a point are represented by a vertical line in the phase space, and the system is required to faithfully map line onto line (Figure 4.4b). That may appear quite demanding, but it is precisely what an imaging system is asked to perform. But suppose we consider only the boundary or edge of all the rays. Then all we require is that the boundary be transported from the source to the target. The interior rays will come along (Figure 4.4c). They cannot "leak out" because were they to cross the boundary, they would first become the boundary, and it is the boundary that is being transported. To complete the analogy, the volume of the container of rays is unchanged in the process. This is the conservation of phase space volume. It is very much like transporting a container of an incompressible fluid—say, water. The fact that elements inside the container mix or the container itself is deformed is of no consequence. To carry the analogy a bit further, suppose one were faced with the task of transporting a vessel (the volume in phase space) filled with alphabet blocks spelling out a message. Then one would have to take care not to shake the container and thereby scramble the blocks. But if one merely needs to transport the blocks without regard to the message, the task is much easier. This is the key idea of nonimaging optics.

4.4   THE EDGE-RAY PRINCIPLE OR "STRING" METHOD

Figure 4.4c suggests the notion of transporting the boundary or edge of the container of rays in phase space. This leads to one of the most useful algorithms of nonimaging optics. We shall see that transporting the edges only, without regard for interior order, allows attainment of the sine law of concentration limit. To motivate the method we begin with the statement that all imaging optical design derives from Fermat's principle. The optical path length between object and image points are the same for all rays (Figure 4.5a). When this same principle is applied to "strings" instead of rays,

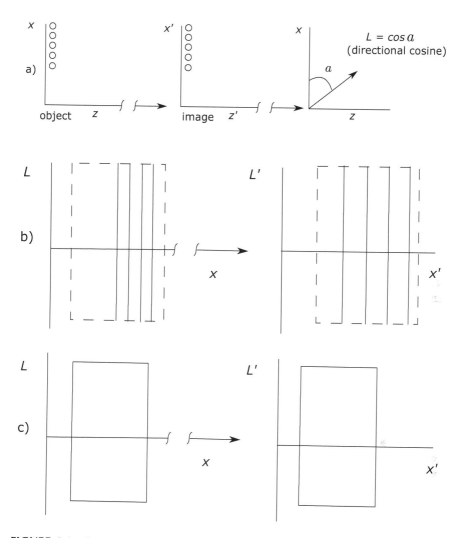

FIGURE 4.4 Comparing imaging with nonimaging in the phase space representation. (a) The spatial representation of source and destination, L, is the direction cosine. (b) The phase space representation of imaging optics. (c) The phase space representation of nonimaging optics (edge-ray method).

it gives the edge-ray algorithm of nonimaging optical design (Figure 4.5b). First we must explain what strings are, which is best done by example. We will proceed to solve the problem of attaining the sine law limit of concentration for the simplest case: that of a flat absorber. Referring to Figure 4.6, we loop one end of a "string" to a "rod" tilted at angle θ to the aperture AA' and tie the other end to the edge of the exit aperture B'. Holding the length fixed, we trace out a reflector profile as the string moves from C to A'. From simple geometry, the relation $BB' = AA' \sin\theta$ immediately follows. This construction gives the 2D compound parabolic concentrator, or CPC. Rotating the profile about the axis of symmetry gives the 3D CPC with radius (a)

Imaging Optics
$\int_{P}^{P'} \underset{rays}{ndl}$ = constant [Fermat 1601-1665], where
n - index of refraction
I = path length
a)
P
P'
b)
P
$\int_{P}^{P'} \underset{strings}{ndl}$ = constant
P'

FIGURE 4.5 Fermat's principle for rays and strings.

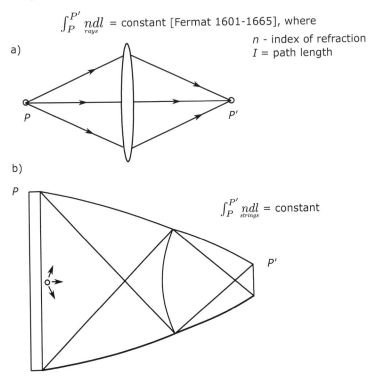

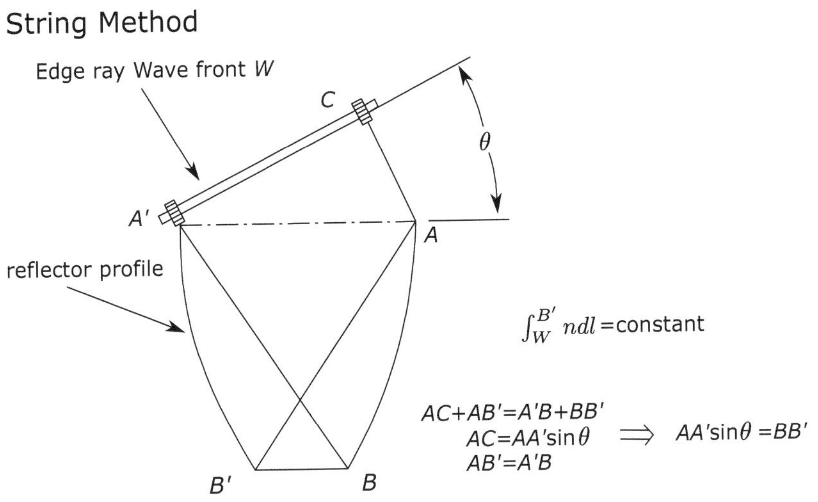

String Method
Edge ray Wave front W
C
θ
A'
A
reflector profile
$\int_{W}^{B'} ndl$ =constant
AC+AB'=A'B+BB'
AC=AA'sinθ
AB'=A'B
⟹ AA'sinθ =BB'
B'
B

FIGURE 4.6 String construction for planar absorber.

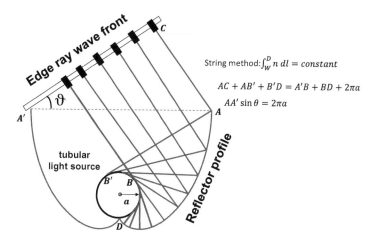

String method: $\int_W^D n\,dl = constant$

$AC + AB' + B'D = A'B + BD + 2\pi a$

$AA' \sin \theta = 2\pi a$

FIGURE 4.7 String construction for tubular absorber.

at the entrance and (b) at the exit. The 2D CPC is an ideal concentrator—that is, it works perfectly for all rays within the acceptance angle q (in 2D geometry). As we will see, the 3D CPC is very close to ideal in 3D geometry. The flat absorber case is a natural candidate for rotating about the axis because the square of the ratio of diameters (sin 2θ) agrees with the maximum concentration. Other absorber shapes such as circular cross-sections (cylinders in 2D, spheres in 3D) do not have this correspondence because the area of the sphere is $4\pi b^2$, whereas the entrance aperture area is πa^2. Notice that we have kept the optical length of the string fixed. For media with varying index of refraction (n), the physical length is multiplied by n. Of course, we have not demonstrated that this construction actually works. One admittedly tongue-in-cheek approach is to state that anything this "neat"—in other words, that satisfies the conservation laws in a natural way—has to work. Perhaps a more serious "proof" is to notice that the 2D CPC rejects all stray radiation and therefore must be ideal for conservation of phase space (Ries & Rabl, 1994; Winston, 1970). The string construction is very versatile and can be applied to any convex (or at least nonconcave) absorber. Figure 4.7 shows the string construction for a tubular absorber as would be appropriate for a solar thermal concentrator.

This time $2\pi a = BB' \sin\theta$, where a is the radius of the cylindrical absorber (Winston & Hinterberger, 1975).

4.5 LIGHT CONES

A primitive form of nonimaging concentrator, the light cone, has been used for many years (see, e.g., Holter et al., 1962). Figure 4.8 shows the principle. If the cone has semiangle γ and if θ_i is the extreme input angle, then the ray indicated will just pass after one reflection if $2\gamma = \left(\dfrac{\theta}{2}\right) - \theta_i$. It is easy to arrive at an expression for the length of the cone for a given entry aperture diameter. Also, it is easy to see that some other rays incident at angle θ_i, such as that indicated by the double arrow,

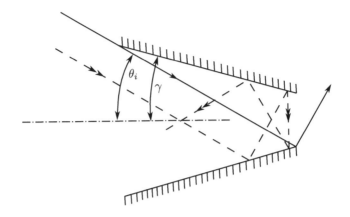

FIGURE 4.8   The cone concentrator.

will be turned back by the cone. If we use a longer cone with more reflections, we still find some rays at angle θ_i being turned back. Clearly, the cone is far from being an ideal concentrator. Williamson (1952) and Witte (1965) attempted some analysis of the cone concentrator but both restricted this treatment to meridian rays. This unfortunately gives a very optimistic estimate of the concentration. Nevertheless, the cone is very simple compared to the image-forming concentrators described in Chapter 3 and its general form suggests a new direction in which to look for better concentrators.

4.6   THE COMPOUND PARABOLIC CONCENTRATOR

The flat absorber case occupies a special place because of its simplicity. Historically it was the first to be discovered. For these reasons its description and properties merit a separate discussion.

If we attempt to improve on the cone concentrator by applying the edge-ray principle, we arrive at the compound parabolic concentrator (CPC), the prototype of a series of nonimaging concentrators that approach very close to being ideal and having the maximum theoretical concentration ratio. Descriptions of the CPC appeared in the literature in the mid-1960s in widely different contexts. The CPC was described as a collector for light from Cerenkov counters by Hinterberger and Winston (1976a,b). Baranov (1965) a paper in Russian that introduces certain propeties of CPC's, a suprising similarity is seen between Baranov and Melnikov (1966) and Hinterberger and Winston (1966). It takes a second look to realize the transmission curves in Figure 4.14 are derived from raytracing while the tranmisssion patterns in Baranov and Melnikoff are obtained photographically. Almost simultaneously, Baranov (1965) and Baranov and Mel'nikov (1966) described the same principle in 3D geometry, and Baranov (1966) suggested 3D CPCs for solar energy collection. Baranov (1965, 1967) obtained Soviet patents on several CPC configurations. Axially symmetric CPCs were described by Ploke (1967), with generalizations to designs incorporating refracting elements in addition to the light-guiding reflecting wall. Ploke (1969) obtained a German patent for various photometric applications.

In other applications to light collection for applications in high-energy physics, Hinterberger and Winston (1966a,b; 1968a,b) noted the limitation to $1/\sin 2\theta$ of the attainable concentration, but it was not until some time later that the theory was given explicitly (Winston, 1970). In the latter publication the author derived the generalized étendue (see appendix A) and showed how the CPC approaches closely to the theoretical maximum concentration.

The CPC in 2D geometry was described by Winston (1974). Further elaborations may be found in Winston and Hinterberger (1975) and Rabl and Winston (1976). Applications of the CPC in 3D form to infrared collection (Harper et al., 1976) and to retinal structure (Baylor & Fettiplace, 1975; Levi-Setti et al., 1975; Winston & Enoch, 1971) have also been described. The general principles of CPC design in 2D geometry are given in a number of U.S. patents (Winston, 1975, 1976a, 1977a,b).

Let us now apply the edge-ray principle to improve the cone concentrator. looking at Figure 4.9, we require that all rays entering at the extreme collecting angle θ_i emerge through the rim point P' of the exit aperture. If we restrict ourselves to rays in the meridian section, the solution is trivial, since it is well known that a parabolic shape with its axis parallel to the direction θ_i and its focus at P' will do this, as shown in Figure 4.10. The complete concentrator must have an axis of symmetry if it is to be a 3D system, so the reflecting surface is obtained by rotating the parabola about the concentrator axis (not about the axis of the parabola).

The symmetry determines the overall length. In the diagram the two rays are the extreme rays of the beam at θ_i, so the length of the concentrator must be such as to just pass both these rays. These considerations determine the shape of the CPC completely in terms of the diameter of the exit aperture $2a'$ and the maximum input angle θ_i. It is a matter of simple coordinate geometry (Appendix C) to show that the focal length of the parabola is

$$f = \frac{a'}{1 + \sin \theta_i} \tag{4.1}$$

that the overall length is

$$L = \left(a'\left(1 + \sin \theta_i\right)\cos \theta_i\right) / \sin^2 \theta_i \tag{4.2}$$

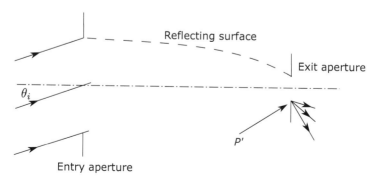

FIGURE 4.9   The edge-ray principle.

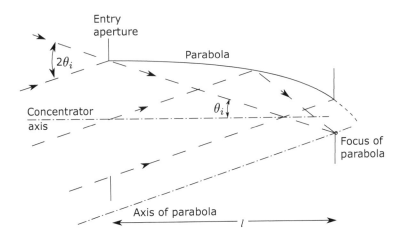

FIGURE 4.10 Construction of the CPC profile from the edge-ray principle.

and that the diameter of the entry aperture is

$$a = \frac{a'}{\sin \theta_i} \tag{4.3}$$

Also, from Equations (4.2) and (4.3) or directly from the figure,

$$L = \left(a + a'\right)\cot \theta_i \tag{4.4}$$

Figure 4.11 shows scale drawing of typical CPCs with a range of collecting angles. It is shown in Appendix C that the concentrator wall has zero slope at the entry aperture, as drawn.

The most remarkable result is Equation (4.3). We see from this that the CPC would have the maximum theoretical concentration ratio (see Section 2.7)

$$\frac{a}{a'} = \frac{1}{\sin \theta_i} \tag{4.5}$$

provided all the rays inside the collecting angle θ_i actually emerge from the exit aperture. Our use of the edge-ray principle suggests that this ought to be the case, given the analogy with image-forming concentrators, but in fact this is not so. The 3D CPC, like the cone concentrator, has multiple reflections, and these can actually turn back the rays that enter inside the maximum collecting angle. Nevertheless, the transmission-angle curves for CPCs, as calculated by ray-tracing, approach very closely the ideal square shape, of course in practical implementation we usually shorten the length by ~1/3 (called truncation). Figure 4.12, after Winston (1970), shows a typical transmission-angle curve for a CPC with $\theta_i = 16°$. It can be seen that the CPC comes very close to being an ideal concentrator. Also, it has the advantages of being a very practical design, easy to make for all wavelengths, since it depends on reflection rather than refraction, and of not requiring any extreme material properties. The only disadvantage is that it is very long compared to its diameter, as can be seen from Equation (4.2). This can

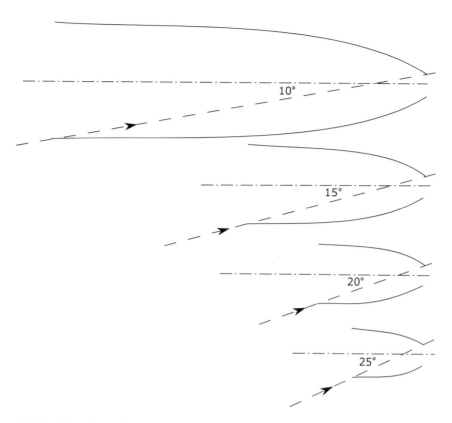

FIGURE 4.11 Some CPCs with different collecting angles. The drawings are to scale with the exit apertures all equal in diameter.

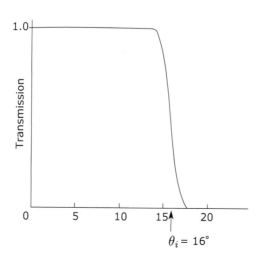

FIGURE 4.12 Transmission-angle curve for a CPC with acceptance angle $\theta_i = 16°$. The cut-off occurs over a range of about 1°.

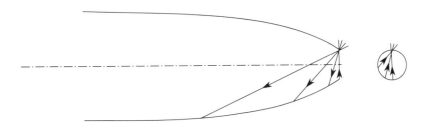

FIGURE 4.13 Identifying rays that are just turned back by a cone like concentrator. The rays shown are intended as projections of skew rays, since the meridional rays through the rim correspond exactly to θ_i by construction for a CPC.

be overcome if we incorporate refracting elements into the basic design. In later sections of this chapter we shall study the optics of the CPC in detail. We shall elucidate the mechanism by rays inside the collecting angle which are turned back, give transmission-angle curves for several collecting angles, and give quantitative comparisons with some of the other concentrators, imaging and nonimaging, that have been proposed. In later chapters we shall discuss modifications of the basic CPC along various lines—for example, incorporating transparent refracting materials into the design and even making use of total internal reflection at the walls for all the accepted rays.

We conclude this section by examining the special case of the 2D CPC or trough like a concentrator. This has great practical importance in solar energy applications, since, unlike other trough collectors, it does not require diurnal guiding to follow the sun. The surprising result is obtained that the 2D CPC is actually an ideal concentrator of maximum theoretical concentration ratio—that is, no rays inside the maximum collecting angle are turned back. To show this result we have to find a way of identifying rays that do get turned back after some number of internal reflections. The following procedure for identifying such rays actually applies not only to CPCs but to all axisymmetric cone-like concentrators with internal reflections. It is a way of finding rays on the boundary between sets of rays that are turned back and rays that are transmitted. These extreme rays must just graze the edge of the exit aperture, as in Figure 4.13, so that if we trace rays in reverse from this point in all directions as indicated, these rays appear in the entry aperture on the boundary of the required region. Thus, we could choose a certain input direction, find the reverse traced rays having this direction, and plot their intersections with the plane of the input aperture. They could be sorted according to the number of reflections involved and the boundaries plotted out. Diagrams of this kind will be given for 3D CPCs in the next chapter.

Returning to the 2D CPC, we note first that the ray-tracing in any 2D trough like a reflector is simple even for rays not in a plane perpendicular to the length of the trough. This is because the normal to the surface has no component parallel to the length of the trough, and thus the law of reflection, Equation(2.1), can be applied in two dimensions only. The ray direction cosine in the third dimension is constant. Thus, if Figure 4.14 shows a 2D CPC with the length of the trough perpendicular to the plane of the diagram, all rays can be traced using only their projections on

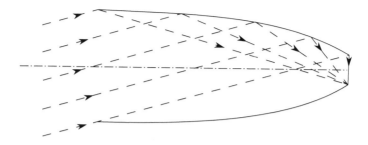

FIGURE 4.14   A 2D CPC. The rays drawn represent projections of rays out of the plane of the diagram.

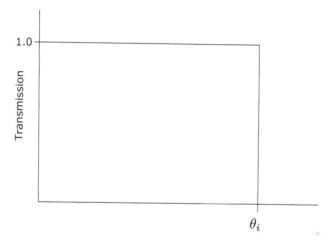

FIGURE 4.15   The transmission-angle curve for a 2D CPC.

this plane. We can now apply our identification of rays that get turned back. Since, according to the design, all the rays shown appear in the entry aperture at θ_{max}, there can be no returned rays within this angle. The 2D CPC has the maximum theoretical concentration ratio and its transmission-angle graph therefore has the ideal shape, as in Figure 4.15.*

Since this property is of prime importance, we shall examine the ray paths in more detail to strengthen the verification. Figure 4.16 shows a 2D CPC with a typical ray at the extreme entry angle θ_{max}. Say this ray meets the CPC surface at P. A neighboring ray at a smaller angle would be represented by the broken line. There are then two possibilities. Either this ray is transmitted as in the diagram, or else it meets the surface again at $P1$. In the latter case we apply the same argument except using the

---

* Strictly, this applies to 2D CPCs that are indefinitely extended along the length of the trough. In practice, this effect is achieved by closing the ends with plane mirrors perpendicular to the straight generators of the trough. This ensures that all rays entering the rectangular entry aperture within the acceptance angle emerge from the exit aperture.

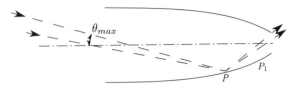

FIGURE 4.16   To prove that a 2D CPC has an ideal transmission-angle characteristic.

extreme ray incident at $P1$, and so on. Thus, although some rays have a very large number of reflections, eventually they emerge if they entered inside θ_{max}. Of course, in the preceding argument "ray" includes "projection" of a ray skew to the diagram.

This result shows a difference between 2D and 3D CPCs. The 2D CPC has maximum theoretical concentration, in the sense of Section 2.9. In extending it to 3D, however, we have included more rays (there is now a threefold infinity of rays, allowing for the axial symmetry, whereas in the 2D case we have to consider only a twofold infinity). We have no more degrees of freedom in the design, since the 3D concentrator is obtained from the 2D profile by rotation about the axis of symmetry.

The 3D concentrator has to be a figure of revolution, and thus we can do nothing to ensure that rays outside the meridian sections are properly treated. We shall see in Section 4.7.3 that it is the rays in these regions that are turned back by multiple reflections inside the CPC.

This discussion also shows the different causes of nonideal performance of imaging and nonimaging systems. The rays in an image-forming concentrator such as a high-aperture lens all pass through each surface the same number of times (usually once), and the nonideal performance is caused by geometrical aberrations in the classical sense. In a CPC, on the other hand, different rays have different numbers of reflections before they emerge (or not) at the exit aperture. It is the effect of the reflections in turning back the rays that produces nonideal performance.

Thus, there is an essential difference between a lens with large aberrations and a CPC or other nonimaging concentrator. A CPC is a system of rotational symmetry, and it would be possible to consider all rays as having just, say, three reflections and to discuss the aberrations (no doubt very large) of the image formation by these rays. But there seems no sense in which rays with different numbers of reflections could be said to form an image. It is for this reason that we continue to draw the distinction between image-forming and nonimaging concentrators.

4.7   PROPERTIES OF THE COMPOUND PARABOLIC CONCENTRATOR

In this section we examine the properties of the basic CPC, the design of which was developed in the last section. We'll see how ray-tracing can be done, the results of ray-tracing in the form of transmission-angle curves, certain general properties of these curves, and the patterns of rays in the entry aperture that get turned back. This detailed examination will help in elucidating the mode of action of CPCs and their derivatives, to be described in later chapters.

4.7.1 THE EQUATION OF THE CPC

By rotation of axes and translation of origin we can write down the equation of the meridian section of a CPC. In terms of the diameter $2a'$ of the exit aperture and the acceptance angle $\theta_{\max}$ this equation is

$$\left(r\cos\theta_{\max} + z\sin\theta_{\max} \right)^2 + 2a'\left(1+\sin\theta_{\max} \right)^2 r$$

$$- 2a'\cos\theta_{\max}\left(2+\sin\theta_{\max} \right)^2 z \qquad (4.6)$$

$$- a'^2\left(1+\sin\theta_{\max} \right)\left(3+\sin\theta_{\max} \right) = 0$$

where the coordinates are as in Figure 4.17. Recalling that the CPC is a surface of revolution about the z axis we see that in three dimensions, with $r^2 = x^2 + y^2$, Equation (4.6) represents a fourth-degree surface.

A more compact parametric form can be found by making use of the polar equation of the parabola. Figure 4.18 shows how the angle ϕ is defined. In terms of this angle and the same coordinates (r, z) the meridian section is given by

$$r = \frac{2f\sin\left(\phi - \theta_{\max}\right)}{1-\cos\phi} - a', \quad z = \frac{2f\cos\left(\phi - \theta_{\max}\right)}{1-\cos\phi} \qquad (4.7)$$

where $f = a'\left(1+\sin\theta_{\max} \right)$

If we introduce an azimuthal angle Ψ we obtain the complete parametric equations of the surface:

$$x = \frac{2f\sin\Psi\sin\left(\phi - \theta_{\max}\right)}{1-\cos\phi} - a'\sin\Psi$$

$$y = \frac{2f\cos\Psi\sin\left(\phi - \theta_{\max}\right)}{1-\cos\phi} - a'\cos\Psi$$

$$z = \frac{2f\cos\left(\phi - \theta_{\max}\right)}{1-\cos\phi} \qquad (4.8)$$

The derivations of these equations are sketched in Appendix C.

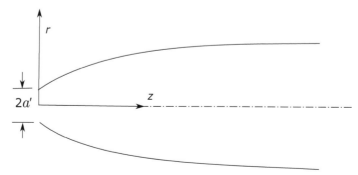

FIGURE 4.17   The coordinate system for the $r - z$ equation for the CPC.

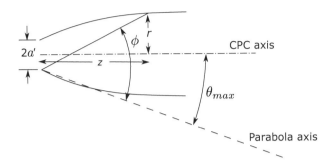

FIGURE 4.18   The angle ϕ used in the parameteric equations of the CPC.

4.7.2   THE NORMAL TO THE SURFACE

We need the direction cosines of the normal to the surface of the CPC for ray-tracing purposes. There are well-known formulas of differential geometry that give these. If the explicit substitution $r = \left(x^2 + y^2 \right)^{\frac{1}{2}}$ is made in Equation (4.6), and the result is written in the form

$$F\left(x, y, z\right) = 0 \tag{4.9}$$

then the direction cosines are given by

$$n = \left(F_x, F_y, F_z \right) / \left(F_x^2 + F_y^2 + F_z^2 \right)^{\frac{1}{2}} \tag{4.10}$$

The formulas for the normal are slightly more complicated for the parametric form. We first define the two vectors

$$a = \left(\partial x / \partial \phi, \partial y / \partial \phi, \partial z / \partial \phi \right), \ \ b = \left(\partial x / \partial \Psi, \partial y / \partial \Psi, \partial z / \partial \Psi \right), \tag{4.11}$$

and then the normal is given by

$$n = a \times b / \left\{ \left| a \right|^2 \left| b \right|^2 - \left| a \cdot b \right|^2 \right\}^{\frac{1}{2}} \tag{4.12}$$

These results are given in elementary texts such as Weatherburn (1931). Although the formulas for the normal are somewhat opaque, it can be seen from the construction for the CPC profile in Figure 4.10 that at the entry end the normal is perpendicular to the CPC axis—that is, the wall is tangent to a cylinder.

4.7.3   TRANSMISSION-ANGLE CURVES FOR CPCs

In order to compute the transmission properties of a CPC, the entry aperture was divided into a grid with spacing equal to 1/100 of the diameter of the aperture and rays were traced at a chosen collecting angle θ at each grid point. The proportion of these rays that were transmitted by the CPC gave the transmission $T(\theta, \theta_{max})$ for the CPC with maximum collecting angle θ_{max}. $T(\theta, \theta_{max})$ was then plotted against θ

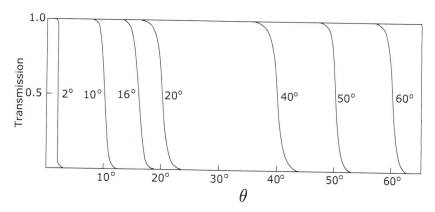

FIGURE 4.19 Transmission-angle curves for 3D CPCs with θ_{max} from 2° to 60°.

to give the transmission-angle curve. Some of these curves are given in Figure 4.19. They all approach very closely the ideal rectangular cutoff that a concentrator with maximum theoretical concentration ratio should have. The transition from $T=0.9$ to $T=0.1$ takes place in $\Delta\theta$ less than 3° in all cases. Approximate values are:

θ_{max}	2°	10°	16°	20°	40°	60°
$\Delta\theta$	0.4°	1.5°	2°	2.5°	2.7°	2.0°

We may also be interested in the total flux transmitted inside the design collecting angle θ_{max}. This is clearly proportional to

$$\int_{0}^{\theta_{max}} T(\theta, \theta_{max}) \sin 2\theta \, d\theta \qquad (4.13)$$

and if we divide by, we obtain Equation 4.13, the fraction transmitted of the flux incident inside a cone of semi-angle θ_{max}. The result of such a calculation is shown in Figure 4.20. This gives the proportion by which the CPC fails to have the theoretical maximum concentration ratio. For example, the 10° CPC should have the theoretical concentration ratio $cosec^2\ 10° = 33.2$, but from the graph it will actually have 32.1. The loss is, of course, because some of the skew rays have been turned back by multiple reflections inside the CPC.

It is of some considerable theoretical interest to see how these failures occur. By tracing rays at a fixed angle of incidence, regions could be plotted in the entry aperture showing what happened to rays in each region. Thus, Figure 4.21 shows these plots for a CPC with $\theta_{max} = 10°$ for rays at 8°, 9°, 9.5°, 10°, 20.5°, 11°, 11.5°. Rays incident in regions labeled 0, 1, 2, ... are transmitted by the CPC after zero, one, two, ... reflections; F2, F3, ... indicate that rays incident in those regions begin to turn back after two, three, ... reflections. Rays in the blank regions will still be traveling toward the exit aperture after five reflections. The calculations were abandoned here to save computer time. In the computation of $T(\theta, \theta_{max})$ where these rays were

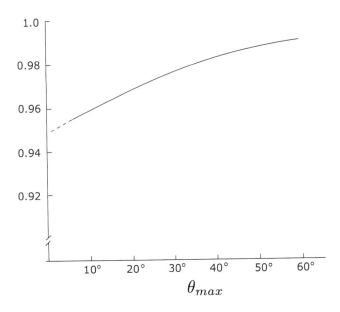

FIGURE 4.20   Total transmission within $\theta_{\max}$ for 3D CPCs.

omitted for θ less than $\theta_{\max}$ it is most likely that all these rays are between the regions for transmission after one and two reflections. This confirms the principle stated in Section 4.3 that rays that meet the rim of the exit aperture are at the boundaries of failure regions. Naturally enough, each split between regions for transmission after n and $n+1$ reflections produces two failure regions, for failure after $n+1$ and $n+2$ reflections.

We can delineate these regions in another way, by tracing rays in reverse from the exit aperture. Thus, in Figure 4.22 we can trace rays in the plane of the exit aperture from a point P at angles γ to the diameter P'. Each ray will eventually emerge from the entry face at a certain angle $\theta(\gamma)$ to the axis and after n reflections. The point in the entry aperture from which this ray emerges is then the point in diagrams, such as those of Figure 4.21, at which the split between rays transmitted after $n-1$ and n reflections begins. For example, to find the points A and B in the 9° diagram of Figure 4.21, we look for an angle γ that yields $\theta(\gamma)=9°$ and find the coordinates of the ray emerging from the entry face after two reflections.

There will, of course, be two such values of γ, corresponding to the two points A and B. This was verified by ray-tracing.

Returning to the blank regions in Figure 4.21, the rays entering at these regions are almost tangential to the surface of the CPC. Thus, they will follow a spiral path down the CPC with many reflections, as indicated in Figure 4.23.

We can use the skew invariant h explained in Section 2.8 to show that such rays must be transmitted if the incident angle is less than $\theta_{\max}$. For if we use the reversed rays and take a ray with $\gamma=\pi/2$, as in Figure 4.22, then this ray has $h=a'$. When it has spiraled back to the entry end (with an infinite number of reflections!), it must have the same $h=a'=a\sin\theta_{\max}$—that is, it emerges tangent to the CPC surface at the

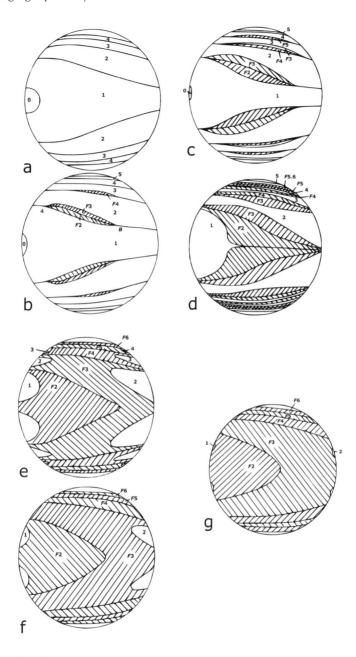

FIGURE 4.21 Patterns of accepted and rejected rays at the entry face of a $10°$ CPC. The entry aperture is seen from above with incident rays sloping downward to the right. Rays entering areas labeled n are transmitted after n reflections; those entering hatched areas labeled Fm are turned back after m reflections. The ray-trace was not carried out to completion in the unlabeled areas. (a) $8°$, $\theta_{max} = 10°$; (b) $9°$, $\theta_{max} = 10°$; (c) $9.5°$, $\theta_{max} = 10°$; (d) $10°$, $\theta_{max} = 10°$; (e) $10.5°$, $\theta_{max} = 10°$; (f) $11°$, $\theta_{max} = 10°$; (g) $11.5°$, $\theta_{max} = 10°$.

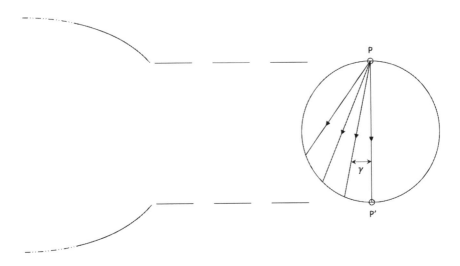

FIGURE 4.22   Rays at the exit aperture used to determine failure regions.

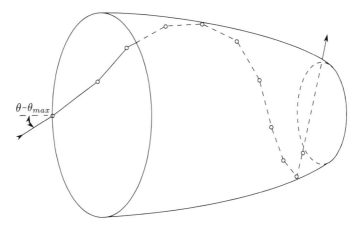

FIGURE 4.23   Path of a ray striking the surface of a CPC almost tangentially.

maximum collecting angle. Any other ray in the blank regions closer to the axis or with smaller θ has a smaller skew invariant and is therefore transmitted.

The preceding argument holds for regions very close to the rim of the CPC.

The reverse ray-tracing procedure shows that for angles below θ_{max} the failures begin well away from the blank region—in fact, at approximately half the radius at the entry aperture. Thus, we are justified in including the blank regions in the count of rays passed for $\theta < \theta_{max}$, as suggested above. This argument also shows that the transmission-angle curves (Figure 4.19) are precisely horizontal out to a few degrees below θ_{max}.

A converse argument shows, on the other hand, that rays incident in this region at angles above θ_{max} will not be transmitted. There seems to be no general argument to show whether the transmission goes precisely to zero at angles sufficiently greater than qmax, but the ray-tracing results suggest very strongly that it does.

4.8 CONES AND PARABOLOIDS AS CONCENTRATORS

Cones are much easier to manufacture than CPCs. Paraboloids of revolution (which of course CPCs are not) seem a more natural choice to conventional optical physicists than concentrators. We shall therefore provide some quantitative comparisons.

It will appear from these that the CPC has very much greater efficiency as a concentrator than either of these other shapes.

In order to make a meaningful comparison, the concentration ratio as defined by the ratio of the entrance and exit aperture areas was made the same as for the CPC with $\theta_{max} = 10°$—that is, a ratio of 5.76 to 1 in diameter. The length of the cone was chosen so that the ray at θ_{max} was just cut off, as in Figure 4.24. For the paraboloid, the exit aperture diameter and the concentration ratio completely determine the shape, as in Figure 4.25.

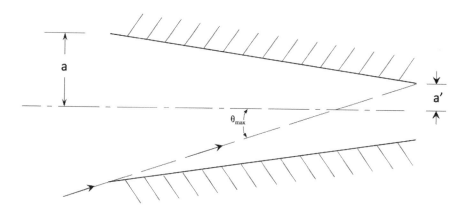

FIGURE 4.24 A cone concentrator, showing dimensions used to compare with a CPC.

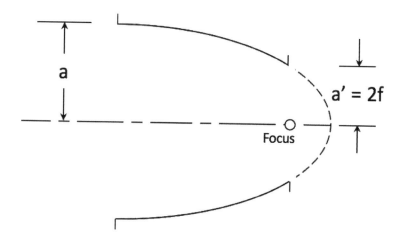

FIGURE 4.25 A paraboloid of revolution as a concentrator.

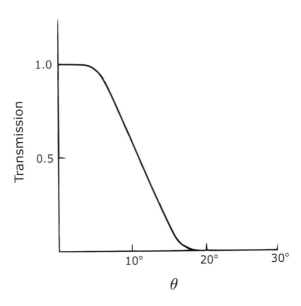

FIGURE 4.26   Transmission-angle curve for a cone; $\theta_{max} = 10°$.

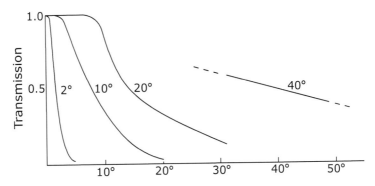

FIGURE 4.27   Transmission-angle curves for paraboloidal mirrors. The graphs are labeled with angles θ_{max} given by $\sin\theta_{max} = a'/a$ in Figure 4.25.

Figures 4.26 and 4.27 show the transmission-angle curves for cones and paraboloids, respectively. It is obvious that the characteristics of both these systems as concentrators are much worse than ideal. For example, the total transmission inside θ_{max} for the paraboloids, according to Equation (4.13), is about 0.60 for all the angles shown. The cones clearly have definitely better characteristics than the paraboloids, with a total transmission inside θ_{max} of order 80%. This is perhaps a verification of our view that nonimaging systems can have better concentration than image-forming systems, since the paraboloid of revolution is an image-forming system, albeit with very large aberrations when used in the present way.*

---

* The section of a paraboloid of revolution in front of the focus is used for x-ray imaging, since most materials are good reflectors for x-rays at grazing incidence.

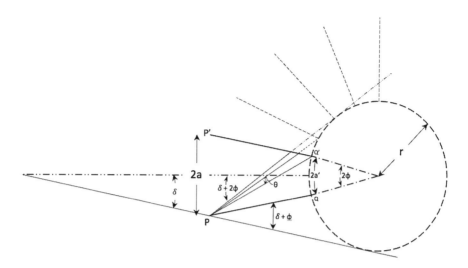

FIGURE 4.28   The pitch-circle construction for reflecting cones and V-troughs. A straight line through the entry aperture of the V-trough emerges from the exit aperture if it cuts the pitch circle. Otherwise, it is turned back. The construction is only valid for meridian rays of a cone.

Rabl (1976b) considered V-troughs—that is, 2D cones—and used a well-known construction (e.g., Williamson, 1952) shown in Figure 4.28 to estimate the angular width of the transition region in the transmission-angle curve. He showed that its width was equal to the angle 2ϕ of the V-trough and the center of the transition came at $\delta + \phi$, where δ is the largest angle of an incident pencil for which all rays are transmitted. If we assume the same holds for a 3D cone, it suggests that the transition in the transmission-angle curve becomes sharper as the cone angle decreases—in other words, smaller-angle cones are more nearly ideal concentrators. This accords with Garwin's result (Garwin, 1952), which may be said to imply that a very long cone with a very small angle is a nearly ideal concentrator.*

A different way to look at the performance of the cone is to note that for small ϕ the concentration ratio a/a' is approximately $1/\sin(\delta + \phi)$. Thus, as the cone length increases while a/a' is held fixed, ϕ/δ tends to zero, and from the diagram, the transition region in the transmission-angle diagram becomes sharper. Nevertheless, there is always a finite transition even for V-troughs and more so for cones so that the comparison with the CPC always shows that the cone is much less efficient and departs further from the ideal than the CPC.

Finally, Figure 4.29 shows the pattern of rays accepted and rejected by a 10° cone as seen at the entry aperture. This may be compared with Figure 4.21d, which shows the pattern for a 10° CPC.

---

* Garwin showed that, in our terminology, a 3D concentrator can be designed to transform an area of any shape into any other while conserving étendue but that it is necessary to start the concentrator as a cylindrical surface and to change its shape adiabatically. In effect, this means that the concentrator would have to be infinitely long to achieve any concentration greater than unity!

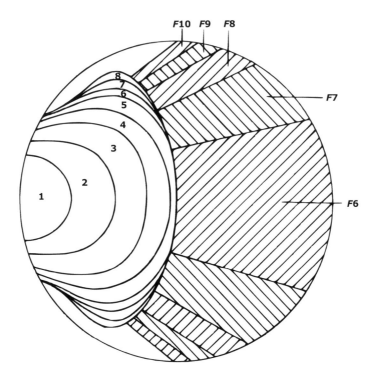

FIGURE 4.29 Patterns of accepted and rejected rays at the entry aperture of a 10° cone for rays at 10°. The notation is as for Figure 4.14, and this figure may be compared directly with Figure 4.14d.

REFERENCES

Baranov, V. K. (1965). A paper in Russian that introduces certain properties of CPCs. *Opt. Mekh. Prom.* **6**, 1–5.

Baranov, V. K. (1966). *Geliotekhnika* **2**, 11–14 [Eng transl.: Parabolotoroidal mirrors as elements of solar energy concentrators. *Appl. Sol. Energy* **2**, 9–12.].

Baranov, V. K. (1967). Device for Restricting in One Plane the Angular Aperture of a Pencil of Rays from a Light Source (in Russian). Russian certificate of authorship 200530, specification published October 31, 1967. Describes certain illumination properties of the 2D CPC, called in other Russian publications a FOCLIN.

Baranov, V. K., and Melnikov, G. K. (1966). Study of the illumination characteristics of hollow focons., *Sov. J. Opt. Technol.* **33**, 408–411.

Baylor, D. A., and Fettiplace, R. (1975). Light and photon capture in turtle receptors. *J. Physiol. (London)* **248**, 433–464.

Brief description of the principle, with photographs to show illumination cutoff and transmission-angle curves plotted from measurements.

Harper, D. A., Hildebrand, R. H., Pernlic, R., and Platt, S. R. (1976). Heat trap: An optimized far infrared field optics system. *Appl. Opt.* **15**, 53–60.

Hinterberger, H., and Winston, R. (1966a). Efficient light coupler for threshold Cerenkov counters. *Rev. Sci. Instrum.* **37**, 1094–1095. See, preprint, Enrico Fermi Institute for Nuclear Studies, EFINS report, 1965.

Hinterberger, H., and Winston, R. (1966b). Gas Cerenkov counter with optimized light-collecting efficiency. *Proc. Int. Conf. Instrum. High Energy Phys.* 205–206. See Also, Navy Large Cyclotron preprint, Enrico Fermi Institute for Nuclear Studies, EFINS report, 1965.

Holter, M. L., Nudelman, S., Suits, G. H., Wolfe, W. L., and Zissis, G. J. (1962). *Fundamentals of Infrared Technology.* Macmillan, New York.

Levi-Setti, R., Park, D. A., and Winston, R. (1975). The corneal cones of *Limulus* as optimized light collectors. *Nature (London)* **253**, 115–116.

Ploke, M. (1967). Lichtführungseinrichtungen mit starker Konzentrationswirkung. *Optik* **25**, 31–43.

Ploke, M. (1969). Axially Symmetrical Light Guide Arrangement. German patent application No. 14722679.

Rabl, A., and Winston, R. (1976). Ideal concentrators for finite sources and restricted exit angles. *Appl. Opt.* **15**, 2880–2883.

Ries, H., and Rabl, A. (1994). Edge-ray principle of nonimaging optics. *J. Opt. Soc. Am. A* **11**, 2627–2632.

Weatherburn, C. E. (1931). *Differential Geometry of Three Dimensions.* Cambridge University Press, London.

Williamson, D. E. (1952). Cone channel condenser optics. *J. Opt. Soc. Am.* **42**, 712–715.

Winston, R. (1970). Light collection within the framework of geometrical optics. *J. Opt. Soc. Am.* **60**, 245–247.

Winston, R. (1974). Principles of solar concentrators of a novel design. *Sol. Energy* **16**, 89–95.

Winston, R. (1976a). Dielectric compound parabolic concentrators. *Appl. Opt.* **15**, 291–292.

Winston, R. (1976b). Radiant Energy Concentration. U.S. letters patent 3923 381.

Winston, R. (1977a). Radiant Energy Concentration. U.S. letters patent 4003 638.

Winston, R. (1977b). Cylindrical Concentrators for Solar Energy. U.S. letters patent 4002 499.

Winston, R., and Hinterberger, H. (1975). Principles of cylindrical concentrators for solar energy. *Sol. Energy* **17**, 255–258.

Witte, W. (1965). Cone channel optics. *Infrared Phys.* **5**, 179–185.

5 Developments and Modifications of the Compound Parabolic Concentrator

5.1 INTRODUCTION

There are several possible ways in which the basic CPC as described in Chapter 4 could be varied for specific purposes. Some of these have been mentioned already, such as a solid dielectric CPC using total internal reflection. Others spring to mind fairly readily for specific purposes, such as collecting from a source at a finite distance rather than at infinity. In this chapter we describe these developments and discuss their properties.

5.2 THE DIELECTRIC-FILLED CPC WITH TOTAL INTERNAL REFLECTION

Both 2D and 3D CPCs filled with dielectric and using total internal reflection were described by Winston (1976). If we consider either the 2D case, or meridian rays in the 3D case, we see that the minimum angle of incidence for rays inside the design-collecting angle occurs at the rim of the exit aperture, as in Figure 5.1. If the dielectric has refractive index n, the CPC is, of course, designed with an acceptance angle θ' inside the dielectric, according to the law of refraction. It is then easy to show that the condition for total internal reflection to occur at all points is

$$\sin \theta'_{max} < 1 - 2/n^2 \text{ or } \sin \theta_{max} < n - (2/n) \tag{5.1}$$

Since the sine function can only take values between 0 and 1, the useful values of n are greater than $\sqrt{2}$. This is in good agreement with the range of useful optical materials in the visible and infrared regions. In a 3D CPC, rays outside the meridian plane have a larger angle of incidence than meridian rays at the same inclination to the axis so Eq. (5.1) covers all CPCs. The expressions in Eq. (5.1) are plotted in Figure 5.2. For most purposes it is unlikely that collecting angles exceeding 40° would be needed, so the range of useful n coincides very well. For trough collectors there is always total internal reflection at the perpendicular end walls, since there is the same angle of incidence here as for the ray at θ' at the entry aperture on the

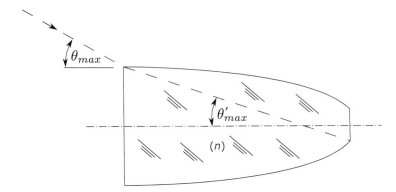

FIGURE 5.1   A dielectric-filled compound parabolic concentrator. The figure is drawn for an sentry angle of 18° and a refractive index of 1.5. The concentration ratio is thus 10.2 for a 3D concentrator.

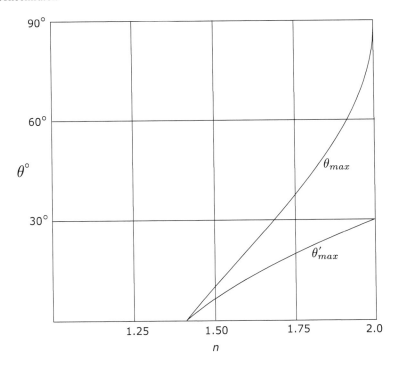

FIGURE 5.2   The maximum collecting angles for a dielectric-filled CPC with total internal reflection, as functions of the refractive index.

curved surface. In fact, it could be shown that for $n > \sqrt{2}$ it is impossible for a ray to get into a trough CPC and not be totally internally reflected at the end face.

The angular acceptance of the dielectric-filled 2D CPC for nonmeridional rays is actually larger than a naive analogy with the $n = 1$ case would indicate. To see this, it is convenient to represent the angular acceptance by the direction cosine variables introduced earlier. Let x be transverse to the trough, let y lie along the trough, and let L and M be the corresponding direction cosines. We recall that the ordinary $(n=1)$ 2D CPC accepts all rays whose projected angles in the x, z plane are $\leq \theta_i$, the design cutoff angle. This condition is represented by an ellipse in the L, M plane with semi-diameters $\sin \theta_i$ and 1, respectively (shown in Figure 5.3a). Therefore, the acceptance

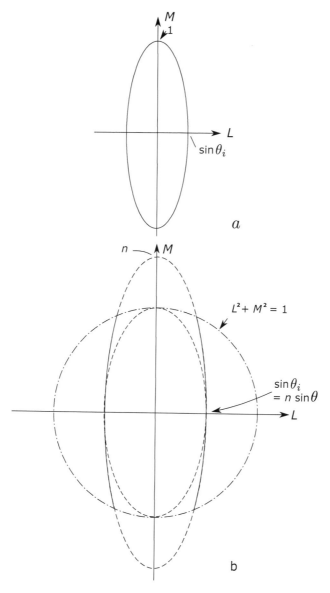

FIGURE 5.3   Angular acceptance for dielectric-filled concentrators, plotted in direction cosine space.

figure just inside the dielectric is such an ellipse with semidiameters $\sin \theta_i$ and a. In terms of direction cosines, Snell's law is simply

$$L = nL', \quad M = nM' \tag{5.2}$$

so one might expect the acceptance ellipse in the L, M plane to be scaled up by n. However, the physical values of L and M lie inside the unit circle

$$L^2 + M^2 < 1 \tag{5.3}$$

It follows that the accepted rays lie inside the intersection of the scaled-up ellipse and the unit circle. This region, as seen in Figure 5.3b, is larger than an ellipse with semi-diameters $\sin \theta_i'$ and 1, respectively. A quantitative measure of this enhancement is useful in discussing the acceptance of such systems for diffuse (i.e., totally isotropic) radiation. Diffuse radiation equal-populates phase space. Hence, it is uniformly distributed in the L, M plane. The area of the acceptance figure for an ordinary 2D CPC (corresponding to Figure 5.3a) is $\pi \sin\theta_i$. Therefore, the fraction of diffuse radiation accepted is just $\sin \theta_i$. However, for the dielectric-filled case, the area depicted by Figure 5.3b is found to be

$$2\left[n \sin \theta_i \tan^{-1}\left(\tan \theta_c \cos \theta_i \right) + \tan^{-1}\left(\tan \theta_i \cos \theta_c \right) \right] \tag{5.4}$$

where θ_c is the critical angle $\sin^{-1}(1/n)$. This exceeds $\pi \sin \theta_i$ by a factor

$$2 / n\left[n \tan^{-1}\left(\tan \theta_c \cos \theta_i \right) + \tan^{-1}\left(\tan \theta_i \cos \theta_c \right) / \sin \theta_i \right] \tag{5.5}$$

This enhancement factor assumes the limiting value for small angles θ_i

$$\frac{2}{n}\left(n\theta_c + \cos \theta_c \right) \tag{5.6}$$

and slowly decreases to unity as θ_i increases to $\pi/2$. For example, for $n=1.5$, a value typical for plastics, the enhancement is a1.17 for small θ_i ($\leq 10°$) and reduces to ~1.13 for $\theta_i = 40°$. We shall return to this property in Chapter 13, where this extra angular acceptance is shown to be advantageous for solar energy collection.

The dielectric-filled CPC has certain practical advantages. Total internal reflection is 100% efficient, whereas it is difficult to get more than about 90% reflectivity from metallized surfaces. Also, for the same overall length the collecting angle in air is larger by the factor n, since Equations (4.1)–(4.4) for the shape of the CPC would be applied with the internal maximum angle θ_i', instead of θ_i. If the absorber can be placed into optical contact with the exit face and if it can utilize rays at all angles of incidence, the maximum theoretical concentration ratio becomes, from Equation (2.13), $n^2/\sin^2\theta_i$—that is, it is increased by the factor n^2 (or n for a trough concentrator).

However, if the rays had to emerge into air at the exit aperture, it is clear that many would get turned back at this interface by total internal reflection. Thus, the CPC design needs to be modified for this case.

5.3   THE CPC WITH EXIT ANGLE LESS THAN $\Pi/2$

There may well be instances such as that mentioned at the end of the last section where it is either impossible or inefficient to use rays emerging at up to $\pi/2$ from the normal to the exit aperture. The CPC designs can be easily modified to achieve this. It would then be close to being what we have called an ideal concentrator but without maximum theoretical concentration ratio.

Let θ_i be the input collecting angle θ_0 and the maximum output angle. Then an ideal concentrator of this kind would, from Equation (2.18), have the concentration ratio

$$C\left(\theta_i,\theta_o\right) = \left(n_0 \sin\theta_o \,/\, n_i \sin\theta_i\right) \tag{5.7}$$

for a 2D system or

$$C\left(\theta_i,\theta_o\right) = \left[\left(\frac{n_0 \sin\theta_o}{n_i \sin\theta_i}\right)\right]^2 \tag{5.8}$$

for a 3D system. Following Rabl and Winston (1976) we may call this device a θ_i/θ_o transformer or concentrator. It is convenient to design the a θ_i/θ_o concentrator by starting at the exit aperture and tracing rays back. As for the basic CPC we start by considering the 2D case or the meridian rays for the 3D case. Let $QQ' = 2a'$ be the exit aperture in Figure 5.4. We make all reversed rays leaving any point on QQ' at angle θ_o appear in the entrance aperture at angle θ_i to the axis. This is easily done by means of a cone section $Q'R$ making an angle $\dfrac{\theta_i - \theta_o}{2}$ with the axis. Next, we make all rays leaving Q at angles less than θ_o appear at the entry aperture at angle θ_i; this is done in the same way as for a CPC by a parabola RP' with focus at Q and axis at angle θ_i to the concentrator axis. The parabola finishes as usual where it meets the extreme ray from Q at θ_i, so its surface is cylindrical at the entry end.

We have ensured by construction that in the meridian section all rays entering at θ_i emerge after one reflection at less than or equal to θ_o, and it is easily seen by

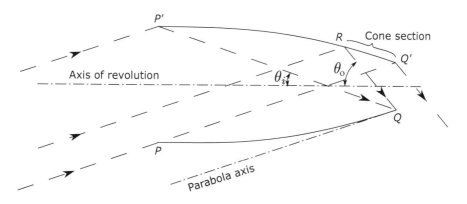

FIGURE 5.4   The $\theta_i\theta_0$ concentrator; as shown, $\theta_i = 18°$ and $\theta_o = 50°$.

examining a few special cases that all rays entering at angles less than θ_i emerge at less than θ_o. This must therefore be an ideal θ_i/θ_o concentrator with

$$a / a' = \sin \theta_o / \sin \theta_i \qquad (5.9)$$

We can also prove this slightly more laboriously from the geometry of the design; this is done in Appendix D.

Clearly the 2D θ_i/θ_o concentrator is an ideal concentrator, since the same reasoning as in Section 4.4 can be applied. In 3D form there will be some losses from skew rays inside θ_i being turned back. We show in Figure 5.5a the transmission-angle curve for a typical case. The transition is as sharp as for a full CPC, but when we compare Figure 5.5b with Figure 4.14, we see that the pattern of rejected rays is quite different.

Although the 2D θ_i/θ_o concentrator transmits within θ_o all rays incident inside θ_i, it does not reject all rays incident outside θ_i but transmits some of them outside θ_o. The reason for this can be seen by noting that the straight cone section in Figure 5.4 lies outside the continuation of the parabolic section, and this parabolic section is the same as the section of an ordinary CPC with entry angle θ_i. Thus, some rays outside θ_i must be transmitted, and since the concentrator is ideal for rays inside θ_i, these extra rays must emerge outside θ_i. The same applies to the 3D version and the effect of these extra rays is not shown in Figure 5.5a.

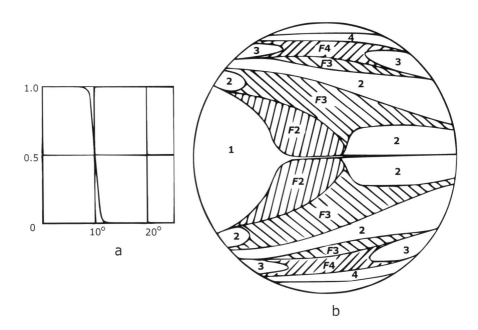

FIGURE 5.5 (a) The transmission-angle curve for a 10°/60° concentrator. (b) Rays transmitted and turned back in the 10°/60° concentrator for 10° input angle (see Figure 4.14 for full legend).

5.4   THE CONCENTRATOR FOR A SOURCE AT A FINITE DISTANCE

So far we have assumed the source to be at infinity, as in the straightforward application to solar energy collection. There are, however, obviously cases where we should like to collect from a source at a finite distance, and the edge-ray principle enables us to derive the shape very simply (Winston, 1978).

In Figure 5.6, let AA' be the finite source and let QQ' be the desired position of the absorber. Then if we apply the edge-ray principle, it is clear that the reflecting surface has the cross section of an ellipse, $P'Q$, with foci at A and Q. For a 3D system the complete surface would be obtained by rotating this ellipse about the axis of symmetry.

We can show that as a 2D system this has maximum theoretical concentration by noting that all rays from AA' that enter the concentrator do emerge (by the same reasoning as for the basic CPC), and then calculating the dimensions of the system by coordinate geometry. This approach is complicated, partly because of the geometry but also because it is not so easy to define the collecting angle for a source at a finite distance. It is better to use a more physical approach and calculate the étendue at either end of the system.

We take an object $AA' = 2\eta$ and an aperture PP' a distance z apart, as in Figure 5.7; if the coordinate y is measured from the center of the aperture, the étendue is

$$\iint d(\sin\theta)\,dy = \int_{-y_{max}}^{y_{max}} \left\{ \frac{(\eta - y)}{\sqrt{z^2 + (\eta - y)^2}} + \frac{(\eta + y)}{\sqrt{z^2 + (\eta + y)^2}} \right\} dy$$

$$= \left[-\sqrt{z^2 + (\eta - y)^2} + \sqrt{z^2 + (\eta + y)^2} \right]_{-y_{max}}^{y_{max}} \qquad (5.10)$$

$$= 2(AP' - AP)$$

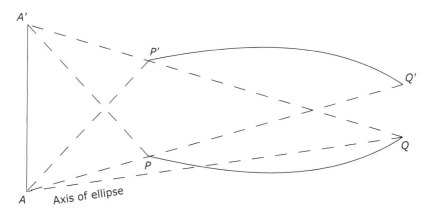

FIGURE 5.6   A concentrator for the object AA' at a finite distance.

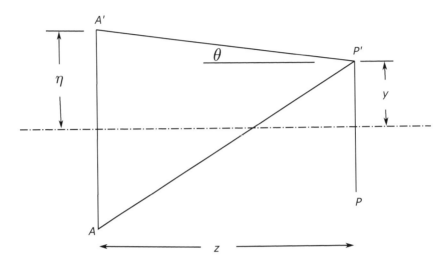

FIGURE 5.7 Calculating the étendue for a source AA' at a finite distance. The collecting aperture is at PP'.

It is useful to rewrite Equation (5.10) in the form

$$(A'P + AP') - (A'P' + AP) \tag{5.11}$$

because this version, a remarkably simple result due to Hottel (1954), actually gives the correct étendue formula even when there is no particular symmetry between the relative positions of aperture and source.

Now, returning to Figure 5.6, we know from the fundamental property of the ellipse that the sum of the distances from the two foci to any point on the curve is a constant—that is,

$$AP + PQ' + Q'Q = AP' + PQ;$$

or

$$AP' - AP = Q'Q \tag{5.12}$$

so that the étendue measured at the output end is $2Q'Q$. Since $Q'Q$ is perpendicular to the axis, this must mean that rays emerge from all points of the exit aperture with their direction cosines distributed uniformly over ±1—that is, this system has the maximum theoretical concentration ratio.

For the 3D case with rotational symmetry (Figure 5.6 now represents a meridional section) a straightforward calculation gives for the étendue (Winston, 1978)

$$\pi^2 (AP' - AP)/4 \tag{5.13}$$

while the maximum value assumed by the skew invariant is

$$h_{\max} = (AP' - AP)/2 \tag{5.14}$$

Notice that, just as for the case of an infinitely distant source (θ_i=constant), both Equations (5.13) and (5.14) are consistent with maximum concentration onto an exit aperture of the diameter given by Equation (5.12). Nevertheless, this system in 3D will turn back some rays, just as for the basic CPC, and so will not be quite ideal.

5.5   THE TWO-STAGE CPC

In Section 5.2 we discussed the dielectric-filled CPC using total internal reflection, and we noted that the use of a refractive index greater than unity at the exit aperture permits in principle a greater concentration ratio. In order to utilize this the absorber must be in optical contact with the dielectric, and preferably also the interface must be matched to minimize reflection losses by a suitable coating or a grading of the refractive index. This seems possible, but we then have the practical difficulty that CPCs are very long and therefore a large volume of possibly expensive dielectric is needed. This difficulty may be circumvented by a two-stage system like the one in Figure 5.8. The first stage in air is a θ_i/θ_o concentrator to give whatever final output angle is needed.

This two-stage system, if designed to give maximum concentration $n/\sin\theta$, where θ is the collecting angle in air, is always longer than a basic CPC designed for concentration $1/\sin\theta$ but with the same collecting aperture. If we make the dielectric part as short as possible—that is, if we maximize θ'' according to Equation (5.1)—then it can be shown that the overall length of the two-stage system is

$$a\left\{\cot\theta + \frac{n}{n^2-2}\cos\theta + \frac{4\left(n^2-1\right)^{\frac{3}{2}}}{n\left(n^2-2\right)^2}\sin\theta\right\} \quad \text{for } \sqrt{2} < n \leq 2$$

$$a\left\{\cot\theta + \cos\theta + \frac{3}{2}\sqrt{3}\sin\theta\right\} \quad \text{for } n > 2 \tag{5.15}$$

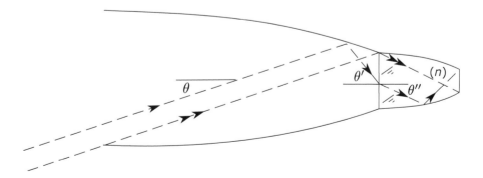

FIGURE 5.8   Two-stage system ending in a dielectric of index n. In the diagram, $\theta = 10°$, $\theta' = 60°$, $\theta'' = 24.5°$, and $n = 1.85$.

whereas the basic CPC has length

$$a\left(\cot\theta + \cos\theta\right) \qquad (5.16)$$

It is easily seen that (5.15) is greater than (5.16). Figure 5.9 shows some typical values for comparison.

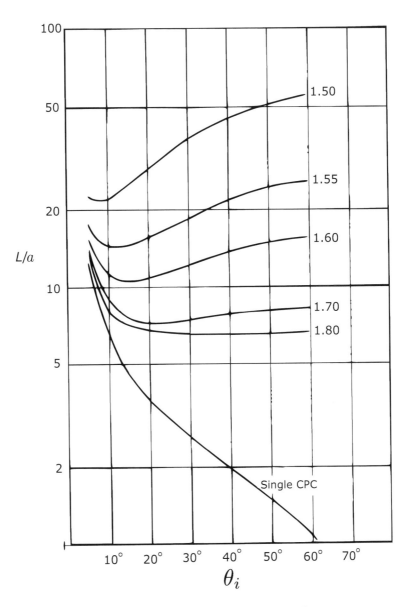

FIGURE 5.9   A comparison of the overall lengths of single- and two-stage concentrators according to Equations (5.15) and (5.16). The ordinate is the ratio of length to radius of collecting aperture, and the graphs are labeled with the refractive index of the second stage.

5.6   THE CPC DESIGNED FOR SKEW RAYS

As we saw in Chapter 4, the basic 3D CPC turns back some skew rays, and this makes its concentration efficiency slightly less than ideal, as shown in Figure 4.15. The basic CPC is designed by applying the edge-ray principle to meridian rays, and this uses up all the degrees of freedom available, so it is not surprising that some skew rays fail. This suggests that we should try designing a concentrator by applying the edge-ray principle to skew rays. The result might then have a slightly closer approach to the maximum theoretical concentration ratio.

Figure 5.10 shows a view from the entry aperture of a concentrator. The entry aperture has a diameter $2a$ and an exit aperture $2a' = 2a\sin\theta_i$. Let this system be designed to fulfill the edge-ray principle for skew rays with skew invariant h (see Section 2.8). In this projected view a ray enters at θ_i and grazes the entry aperture at A, and the exit aperture at B is reflected across to C on the exit aperture (double arrows). The segment BC is thus tangent to a circle of radius $h/\sin\theta_i$. Another ray AD, also entering at θ_i, meets the opposite entry edge at D and is reflected there. In order to satisfy the edge-ray principle, it must meet the edge of the exit aperture and therefore is like the double-arrowed segment AB rotated around the axis so that A reaches D. Thus, the segment reflected from D touches the circle of radius $h/\sin\theta_i$.

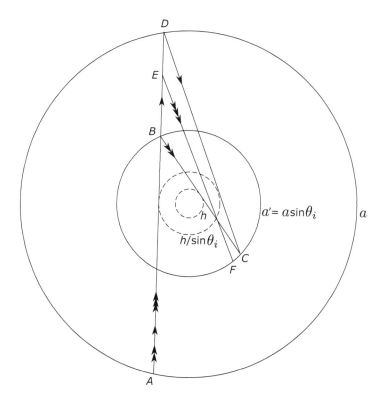

FIGURE 5.10   Design of a concentrator by applying the edge-ray principle to rays with nonzero skew invariant.

We can show that this segment meets the exit aperture at the same point C as the segment BC reflected at B by calculating the angles subtended at the center by the various ray segments.

The design is completed by requiring that all rays that leave the exit aperture at its rim have entered at the input angle θ_i. These rays in the projection of Figure 5.10 are typified by the three-arrowed segments—that is, reflection at some point E along the concentrator to emerge at F on the exit rim. It turns out that the rays do not in general emerge at the same point on the exit rim. Thus, we have a situation where the edge-ray principle is less restrictive than, say, a requirement for imaging at the exit rim. This process completely determines the concentrator as a surface of revolution, but it does not seem to be possible to represent it by any analytical expression. P. Greenman computed the solution for several values of the input angle θ_i and skew invariant h. The results are, briefly, that the shapes are very similar to those of the basic CPC for the same θ_i but that the overall lengths are less and the transitions in the transmission-angle curves are correspondingly more gradual. Figure 5.11 shows some of these curves.

The overall length of this concentrator is determined, as for the basic CPC, by the extreme rays, as in Figure 5.12. This figure shows the rays ADC and ABC of Figure 5.10, both inclined at the extreme input angle to the axis and both grazing the exit aperture after one and no reflections from the concentrator, respectively. Then in order to admit all rays at θ_i or less, the concentrator surface must finish at point A, determined by the intersection of ray ABC with the surface.

This geometry is obvious for the basic CPC, and also in that system the ratio of input to output diameters is set as part of the design data at the desired value $1/\sin \theta_i$. In the present system the design is developed from one end, and it does not follow that the ratio of input to output diameters will have any particular simple value. In fact, it can be shown that this ratio is again $1/\sin \theta_i$—a result that is by no means obvious. In spite of this, the concentrator with nonzero h has even less of the maximum theoretical concentration ratio than the basic CPC does, as can be seen by comparing Figures 5.11 and 4.19. This is because the shorter length permits meridian

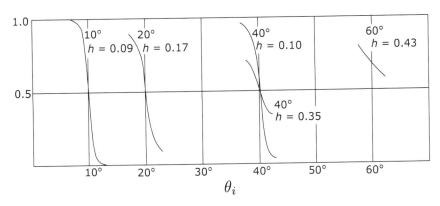

FIGURE 5.11 Transmission-angle curves for concentrators designed for nonzero skew invariant h. All the concentrators have exit apertures of diameter unity.

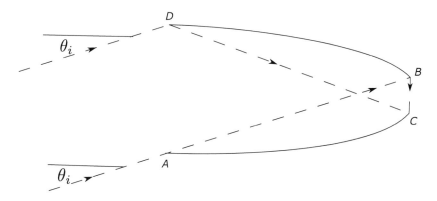

FIGURE 5.12 How the length of a concentrator is determined by the extreme rays.

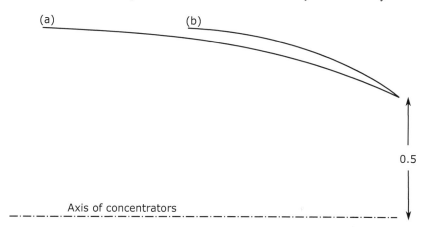

FIGURE 5.13 Comparison of concentrator profiles for $\theta_i = 40°$; (a) $h=0$; (b) $h=0.32$. It can be seen that meridian rays from the edge of (b) at angles greater than 40° are transmitted, since rays at 40° from the edge of (a) just get through the exit aperture.

rays at angles greater than θ_i to reach the exit aperture directly. Thus, by volume conservation of phase space (Appendix A), more rays inside θ_i must be rejected.

Figure 5.13 shows a scale-drawing comparison of the basic 40° CPC with concentrators designed for nonzero h. It can be seen how meridian rays at angles greater than θ_i reach the exit aperture.

5.7   THE TRUNCATED CPC

A disadvantage of the CPC compared to systems with smaller concentration is that it is very long compared to the diameter of the collecting aperture (or width for 2D systems). This is naturally important for economic reasons in large-scale applications such as solar energy. From Equation (4.2) the length L is approximately equal to the diameter of the collecting aperture divided by the full collecting angle—that is,

$$L \sim 2a / 2\theta_i \tag{5.17}$$

If we truncate the CPC by removing part of the entrance aperture end, we find that a considerable reduction in length can be achieved with very little reduction in concentration, so this may be a useful economy.

It is convenient to express the desired relationships in terms of the (r,ϕ) polar coordinate system as in Figure 5.14. We denote truncated quantities by a subscript T. We are interested in the ratio of the length to the collecting apertures and also in the ratio of the area of the reflector to that of the collecting aperture. We find

$$a_T = \frac{f \sin\left(\phi_T - \theta_i\right)}{\sin^2\left(\frac{1}{2}\phi_T\right)} - a' \quad f = a'\left(1 + \sin\theta_i\right)$$

$$a = a' / \sin\theta_i \tag{5.18}$$

$$L_T = \frac{f \cos\left(\phi_T - \theta_i\right)}{\sin^2\left(\frac{1}{2}\phi_T\right)} \tag{5.19}$$

$$L = f \cos\theta_i / \sin^2\theta_i \tag{5.20}$$

so that

$$\frac{L_T}{a_T} = \frac{\left(1 + \sin\theta_i\right)\cos\left(\phi_T - \theta_i\right)}{\sin\left(\phi_T - \theta_i\right)\left(1 + \sin\theta_i\right) - \sin^2\left(\frac{1}{2}\phi_T\right)} \tag{5.21}$$

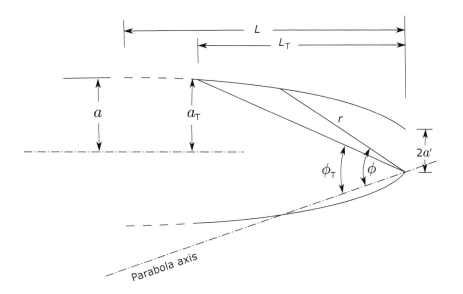

FIGURE 5.14   The polar coordinates used in computing truncation effects.

Plots of this quantity against the theoretical concentration ratio aT/a' for 2D truncated CPCs were given by Rabl (1979) and by Winston and Hinterberger (1975). Figure 5.15 shows some of these curves, and it can be seen that initially—for points near the broken line locus for full CPCs—the curves have a very large slope, so the loss in concentration ratio is quite small for useful truncations.

In addition to the ratio of length to aperture diameter we may be interested in the ratio of the surface area of reflector to aperture area, since this governs the cost of material for the reflector. The general forms of the curves would be similar to those in Figure 5.15 but with differences between 2D and 3D concentrators. The explicit formulae for reflector area divided by collector area are, for a 2D truncated CPC,

$$\frac{-f}{a_T}\left\{\frac{\cos\dfrac{\phi}{2}}{\sin^2\dfrac{\phi}{2}}-\ln\left(\tan\dfrac{\phi}{4}\right)\right\}\Bigg|_{\phi_T}^{\theta_i+\frac{\pi}{2}} \tag{5.22}$$

and for a 3D truncated concentrator

$$\frac{2f}{a_T^2}\int_{\phi_T}^{\theta_i+\frac{\pi}{2}}\left\{\frac{f\sin\left(\phi-\theta_i\right)}{\sin^5\dfrac{\phi}{2}}-\frac{a'}{\sin^3\dfrac{\phi}{2}}\right\}d\phi \tag{5.23}$$

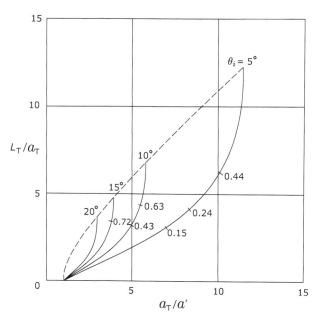

FIGURE 5.15 Length as a function of concentration ratio for 2D truncated concentrators. The numbers marked on the curves are the actual truncation ratios—that is, L_T/L.

Derivations of the preceding results and the explicit form of the integral in Equation (5.23) are given in Appendix E. Some representative plots of these functions are given in Figures 5.16 and 5.17. It should be noted that in these figures the theoretical concentration ratios are respectively (a_T/a') and $(a_T/a')^2$. We conclude from this that losses in performance due to moderate truncation would be acceptable in many instances on account of the economic gains.

5.8   THE LENS-MIRROR CPC

A more fundamental method for overcoming the disadvantage of excessive lengths incorporates refractive elements to converge the pencil of extreme rays. By consistent application of the edge-ray principle we leave the optical properties of the concentrator essentially identical to the all-reflecting counterpart while substantially reducing the length in many cases. The edge-ray principle requires that the extreme incident rays at the entrance aperture also be the extreme rays at the exit aperture. In the all-reflecting construction (Figure 4.10) this is accomplished by a parabolic mirror section that focuses the pencil of extreme rays from the wave front W onto the point $P2$ on the edge of the exit aperture. To incorporate, say, a lens at the entrance aperture, the rays from W, after passage through the lens, are focused onto $P2$ by

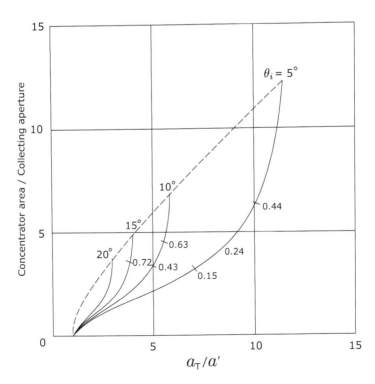

FIGURE 5.16 Concentrator surface area as a function of concentration ratio for 2D truncated concentrators. The numbers marked on the curves are the actual truncation ratios—that is, L_T/L.

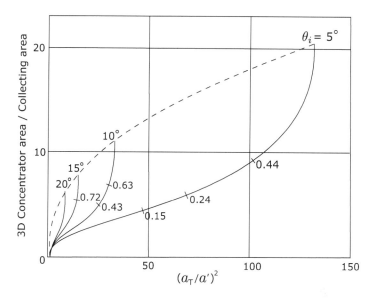

FIGURE 5.17 Concentrator surface area as a function of concentration ratio for 3D truncated concentrators. The numbers marked on the curves are the actual truncation ratios—that is, L_T/L.

an appropriately shaped mirror M (Figure 5.18). Therefore, the profile curve of M is then determined by the condition

$$\int_{W}^{P_2} nds = \text{const} \tag{5.24}$$

To comprehend the properties of the lens-mirror collector, it is useful to consider a hypothetical lens that focuses rays from P onto a point F. From Equation (5.24) the appropriate profile curve for M is a hyperbola with conjugate foci at F and $P2$ (Figure 5.18).* This example illustrates the principal advantage of this configuration. The overall length is greatly reduced from the all-reflecting case to $L=f$, the focal length of the lens. A real lens would have chromatic aberration, so M would no longer be hyperbolic but simply a solution to Equation (5.24). A solution will be possible as long as the aberrations are not so severe as to form a caustic between the lens and the mirror. For the example in Figure 5.18, where the lens is plano-convex with index of refraction $n \sim 1.5$, this means we must not choose too small a value for the focal ratio of this simple lens (an $f/4$ choice works out nicely). Alternatively, we may say that the mirror surface corrects for lens aberrations, providing these are not too severe, to produce a sharp focus at $P2$ for the extreme rays. Of course, this procedure can only be successful for monochromatic aberrations so that it is advantageous to employ a lens material of low dispersion over the wavelength interval of interest.

We may expect the response to skew rays in a rotationally symmetric 3D system to be nonideal just as in the all-reflecting case, and, in fact, ray-tracing of some

---

* By suitable choice of parameters, the hyperbola can be a straight line.

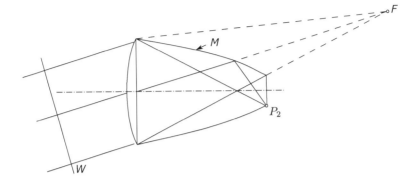

FIGURE 5.18   The lens-mirror CPC.

sample lens-mirror configurations shows angular cutoff characteristics indistinguishable from the simple CPC counterpart. We note that certain configurations of the lens-mirror type were proposed by Ploke (1967).

5.9   2D COLLECTION IN GENERAL

For moderate concentration ratios for solar energy collection, there is considerable interest in systems that do not need diurnal guiding for obvious reasons of economy and simplicity (see, e.g., Winston, 1974; Winston & Hinterberger, 1975). These naturally would have trough-like or 2D shapes and would be set pointing south* at a suitable elevation so as to collect flux efficiently over a good proportion of the daylight hours. So far our discussion has suggested that these might take the form of 2D CPCs, truncated CPCs, or compound systems with a dielectric-filled CPC as the second stage. In discussing all these it was tacitly assumed that the absorber would present a plane surface to the concentrator at the exit aperture, and this, of course, made the geometry particularly simple. In fact, when applications are considered in detail, it becomes apparent that other shapes of absorber would be useful. In particular, it is obvious that cylindrical absorbers—that is, tubes for heating fluids—suggest themselves. In this chapter we discuss the developments in design necessary to take account of such requirements.

5.10   EXTENSION OF THE EDGE-RAY PRINCIPLE

In Chapter 3 we proposed the edge-ray principle as a way of initiating the design of concentrators with concentration ratios approaching the maximum theoretical value. We found that for the 2D CPC this maximum theoretical value was actually attained by direct application of the principle. We now propose a way of generalizing the principle to nonplane absorbers in 2D concentrators. Let the concentrator be as in Figure 5.19, which shows a generalized tubular absorber. We assume the section of

---

* In the northern hemisphere.

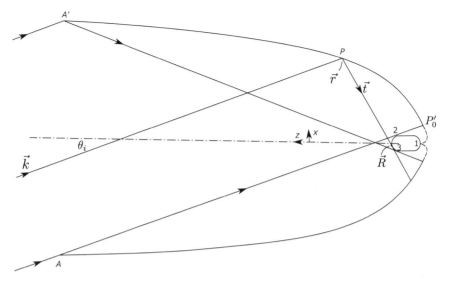

FIGURE 5.19 Generalizing the edge-ray principle for a nonplane absorber.

the absorber is convex everywhere, and we also assume it is symmetric about the horizontal axis indicated.

Then we assert that the required generalization of the edge-ray principle is that rays entering at the maximum angle θ_i shall be tangent to the absorber surface after one reflection, as indicated.

The generalization can easily be seen to reduce to the edge-ray principle for a plane absorber. In order to calculate the concentration we need to have a rule for constructing the concentrator surface beyond the point P_0', at which the extreme reflected ray meets the surface. Here we choose to continue the reflector as an involute of the absorber surface, as indicated by the broken line. A reason for this choice will be suggested in the following. We shall be able to show that this design for a 2D concentrator achieves the maximum possible concentration ratio, defined in this case as the entry aperture area divided by the area of the curved absorber surface.

Following Winston and Hinterberger (1975) we let r be the position vector of a current point P on the concentrator surface and we take R as the position vector of the point of contact of the ray with the absorber. Then we have

$$\vec{r} = \vec{R} - l\vec{t}$$ (5.25)

where t is the unit tangent to the absorber—that is, we have

$$\vec{t} = d\vec{R} / dS$$ (5.26)

where S is the arc length round the absorber. Let k be a unit vector along the direction of the extreme rays so that in the coordinate system shown $k = (\sin \theta_i, 0, -\cos \theta_i)$. Our condition that the tangent to the absorber be reflected into k takes the form, equating sines of the angles of incidence and reflection,

$$\vec{t} \cdot d\vec{r} = \vec{k} \cdot d\vec{r}$$

or

$$\vec{t} \cdot d\vec{r} / dS = \vec{k} \cdot d\vec{r} / dS \qquad (5.27)$$

Now by differentiating Equation (5.25) we obtain

$$d\vec{r} / dS = d\vec{R} / dS - dl / dS\vec{t} - ld\vec{t} / dS$$

and on scalar multiplication by t this gives

$$\vec{t} \cdot d\vec{r} / dS = 1 - dl / dS \qquad (5.28)$$

On substituting in Equation (5.27) and integrating we obtain

$$\left.(S - l)\right|_2^3 = \left(\vec{r_3} - \vec{r_2}\right) \cdot \vec{k} \qquad (5.29)$$

In this equation points 2 and 3 would be those corresponding to the extreme reflected rays, as in the diagram.

Between points 1 and 2 we have postulated that the concentrator profile shall be an involute of the absorber, and the condition for this is

$$\vec{t} \cdot d\vec{r} / dS = 0 \qquad (5.30)$$

Thus, for this section of the curve we have from Equation (5.28)

$$S_2 - S_1 = l_2 - l_1$$

or, since our involute is chosen to be the one that starts at point 1

$$S_2 = l_1 \qquad (5.31)$$

Thus, Equation (5.29) gives

$$S_3 - l_3 = \left(\vec{r_3} - \vec{r_2}\right) \cdot \vec{k} \qquad (5.32)$$

From the figure it can be seen that $\left(\vec{r_3} - \vec{r_2}\right) \cdot \vec{k}$ is equal to the projection of $P_0'A'$ onto AP_0'. Thus

$$\left(\vec{r_3} - \vec{r_2}\right) \cdot \vec{k} = -\left(l_3 + l_2\right) + 2a \sin \theta_i$$

and on substituting into Equation (5.32) we find

$$S_3 + l_2 = 2a \sin \theta_i \qquad (5.33)$$

Recalling that the second section of the concentrator is an involute, we see that $l_2 = S_2$. Thus,

$$S = S_3 + S_2 = 2a \sin \theta_i \qquad (5.34)$$

We have proved that the concentrator profile generated in this way has the theoretical ratio of input area to absorber area—that is, it has the maximum theoretical concentration ratio if no rays are turned back.

If the property of the involute—that its normal is tangent to the parent curve—is remembered, it is easy to see that a concentrator designed in this way sends all rays inside the angle θ_i to the absorber, including those outside the plane of the diagram if it is a 2D system. Thus, from arguments based on étendue and on phase space conservation (see Section 2.7 and Appendix A) the system is optimal.

5.11   SOME EXAMPLES

It is easy to apply our generalization to plane absorbers. Figure 5.20 shows an edge-on fin absorber QQ' with extreme rays AQP_0' and $A'QP_0$. Clearly the section of the concentrator between P_0 and P_0' is an arc of a circle centered on Q and the section $A'P_0'$ is a parabola with focus at Q and axis AQP_0'.

The two-sided flat plate collector normal to the axis, as in Figure 5.21, is a slightly more complicated case. Following our rules, there are three sections to the profile. OP' receives no direct illumination and is thus an involute of the segment OQ'; that is, it is an arc of a circle centered on Q' and therefore part of a parabola with focus at Q and axis $AQ'P'$. $R'A'$ must focus extreme rays on Q and is therefore a parabola with focus Q and axis parallel to $AQ'P'$.

In all cases it can easily be seen from the general mode of construction described in Section 5.10 that the segments of different curves have the same slope where they join; for example, in Figure 5.21 the normal at P' is a ray for the circular segment OP' and the parabolic segment $P'R'$, and at R' the incident ray at angle θ_i must be reflected to Q' by the segment $P'R'$ and to Q by the segment $R'A'$.

Figure 5.22 shows to scale the profile for a circular section absorber. Here the actual profile does not have a simple parabolic or circular shape, but we shall give the solution in Section 5.12. It is noteworthy, however, that in Section 5.10 we deduced

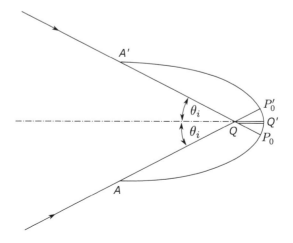

FIGURE 5.20   The optimum concentrator design for an edge-on fin.

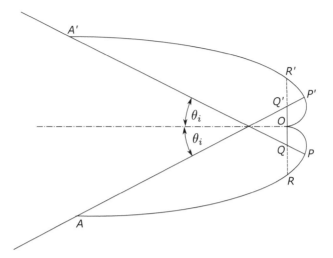

FIGURE 5.21   The optimum concentrator design for transverse fin.

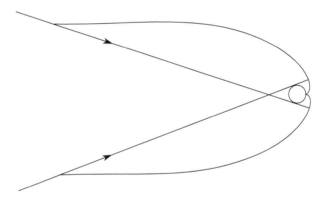

FIGURE 5.22   The optimum concentrator design for a cylindrical absorber.

the property of having the maximum theoretical concentration ratio without explicit reference to the profile, just as we were able to do for the basic CPC (see Chapter 4). For example, Ortobasi (1974) independently developed the ideal mirror profile in an innovative collector program at Corning Glass Company.

5.12   THE DIFFERENTIAL EQUATION FOR THE CONCENTRATOR PROFILE

It is straightforward but laborious to set up a differential equation for the concentrator profile. The equation, given with its solution in Appendix D, is used for the region of the profile that sends extreme rays tangent to the absorber after one reflection—that is, the region between points 2 and 3 in Figure 5.19. The remaining region is an involute arranged to join the profile smoothly, and the equation for this is also given in the appendix. However, it is worth noting that for many practical applications the involute curve can be drawn accurately enough to scale by the draftsman's method of unwinding a taut thread from the absorber profile.

5.13 MECHANICAL CONSTRUCTION FOR 2D CONCENTRATOR PROFILES

In the discussion of Figure 5.19 in Section 5.20, it appeared that part of the concentrator surface was generated as an involute of the absorber section. It is possible to combine this result with the fact that optical path lengths from a wave front to a focus are constant to obtain a simple geometrical construction for the concentrator profile. Figure 5.23 shows a system similar to that shown in Figure 5.19, but we have drawn in a wave front AB of the incoming extreme pencil, and we assume the source is at a large but finite distance. The construction is then as follows. We tie a string between the source and point 1 at the rear of the absorber and pull the string taut with a pencil, as in the so-called gardener's method of drawing an ellipse. The length of the string must be such that it will be just taut when it is pulled right around the absorber to reach point 1 from the other side, as in Figure 5.24. It is then unwound, keeping the string taut, and the pencil describes the correct profile. To check this, we simply have to show that the line drawn is at the correct angle to produce reflection.

In Figure 5.25 the string is tangent to the absorber at A, the source point is at C, and B is a typical position of the pencil. If the pencil is moved to B' where BCB' is a small angle ε, then we have

$$CBA = CB'A' - AA' + O\left(\varepsilon^2\right) \tag{5.35}$$

so that

$$CBA = CB'A + O\left(\varepsilon^2\right) \tag{5.36}$$

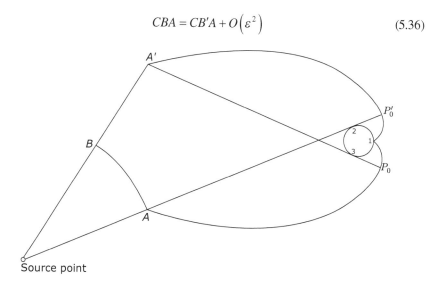

FIGURE 5.23   A concentrator for a source at a finite distance and a nonplane absorber.

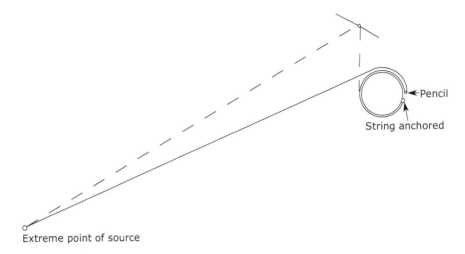

FIGURE 5.24 The string construction for the concentrator profile.

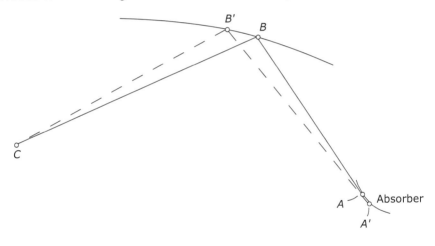

FIGURE 5.25 Proof that the string construction gives a concentrator surface that agrees with Fermat's principle.

Thus, by Fermat's principle BB' must be a portion of a reflecting surface that reflects CB into BA.

Figure 5.26 shows this method generalized further to a convex source and a convex absorber. The string is anchored at two suitably chosen points A and B and stretched with the pencil P. The length of the string is chosen so that it just reaches to point Q when wound around the absorber. It is easily seen that this generates a 2D ideal concentrator. Here we are generating a concentrator that collects all the flux from the source and sends it all to the absorber, and for this to be physically possible, we must make the perimeters of the source and absorber equal. If this were not so, the construction would not work because the string would not be the correct length to just close the curve, and this would mean we were trying to infringe the rules dictated by conservation of étendue.

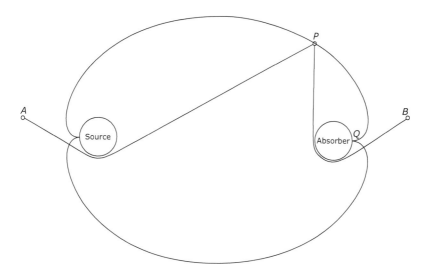

FIGURE 5.26   The case of a convex source and a convex absorber treated by the string construction.

Returning to Figure 5.26 we note that a solution is possible for any distance between source and absorber. We note also that the solution appears not to be unique, in the sense that we could break the reflector at its widest part and insert a straight parallel-sided section of any length, since such a section clearly transforms an étendue of width $2a$ and angle π. However, from obvious practical considerations, it is desirable to minimize the number of reflections of a given ray, and this is clearly done by not inserting such a straight section.

We could, of course, vary these solutions still further by adding in CPCs and θ_0/θ_i systems (Section 5.3) at the preceding break points instead of or in addition to the cylindrical sections. Again, this adds to the number of reflections, and it seems that there is always a unique "most economical" solution for which the extreme rays have at most one reflection.

We could generalize the string method to media with a nonuniform refractive index by postulating that the string is elastic in such a way that it always assumes the optical path length, $\int n\,ds$, of the medium through which it passes. There is, of course, no way in which such a string could be realized, but the concept shows how ray trajectories—that is, geodesics—between the extremes of the apertures always define the correct mirror surface.

In another example of parallel development, the "string method" for generating ideal mirror profiles was discovered independently by Bassett and Derrick (1978) of the University of Sidney.

5.14   A GENERAL DESIGN METHOD FOR A 2D CONCENTRATOR WITH LATERAL REFLECTORS

In this section we formulate a more general treatment of the 2D concentrator with lateral reflectors, from which most other results in this chapter could be derived.

A further generalization will be shown in Chapter 6. We shall describe a procedure in which an input surface and an output surface of given shapes are postulated, enclosing and surrounded by regions of given refractive index distributions.

On each of these surfaces a distribution of extreme rays is given. Then the procedure enables us to design a 2D concentrator that will ensure that all rays between the extreme incoming rays and none outside are transmitted so that the concentrator is optimal.

Suppose we have, as in Figure 5.27, two surfaces AB and $A'B'$, and let AB be illuminated in such a way that the extreme angle rays at each point form pencils belonging respectively to wave fronts Σ_a and Σ_b. Similarly, rays at intermediate angles belong to other wave fronts, so the whole ensemble of rays comes ultimately from a line of point sources and is transformed by a possibly inhomogeneous medium in such a way that the rays just fill the aperture AB. These rays then have a certain étendue H, and we shall see following how to calculate it. Similarly, we draw rays and wave fronts emerging from $A'B'$ as indicated, and we postulate that these shall have the same étendue H.

Now we want to know how we can design a concentrator system between the surfaces AB and $A'B'$, possibly containing an inhomogeneous medium, that shall transform the incoming beam into the emergent beam without loss of étendue.

To solve this problem, we postulate a new principle (we shall see that our edge-ray principle of Chapter 4 can be regarded as being derived from it): the optical system between AB and AB' must be such that it exactly images the pencil from the wave front Σ_α into one of the emergent wave fronts and Σ_β into the other. By "exactly" we mean that all rays from Σ_α as delimited by the aperture AB must just fill $A'B'$ so that none is lost and there is no unused space, and the same for the rays from Σ_β. This principle is relaxed in the formulation of Chapter 6, where it is only required that the optical system image the rays of $\Sigma_\alpha \cup \Sigma_\beta$ into rays of any of the emergent wave fronts.

At this point it may be objected that the preceding seems to have little connection with our original edge-ray principle. But consider a system such as that in Figure 5.28,

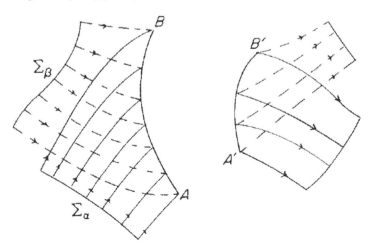

FIGURE 5.27   Beams of equal étendue to fit a concentrator.

which shows rays from one extreme wave front Σ_α in a CPC-like concentrator, and also a wave front Σ'_α gradually moving into coincidence with the edge A' of the exit aperture. We then recover the CPC geometry and the original edge-ray principle. Thus, this new principle could be stated in the form that extreme points of the source must be imaged through the system by rays that just fill the exit and entry apertures.

To see how this principle leads to a solution of the general problem stated at the beginning of this section, we must first show how to calculate the étendue of an arbitrary beam of rays at a curved aperture, as in Figure 5.27. We use the Hilbert integral, a concept from the calculus of variations. In the optics context (Luneburg, 1964) the Hilbert integral for a path from $P1$ to $P2$ across a pencil of rays that originated in a single point is

$$I(1,2) = \int_{P_1}^{P_2} n\vec{k} \cdot ds \qquad (5.37)$$

where n is the local refractive index, k is a unit vector along the ray direction at the current point, and ds is an element along the path $P_1 P_2$. Thus, $I(1, 2)$ is simply the optical path length along any ray between the wave fronts that pass through P_1 and P_2, so it is independent of the form of the path of integration. We can now use this to find the étendue of the beams in Figure 5.29. The Hilbert integral from A to B for the a pencil is seen from Equation (6.14) to be

$$I_\alpha(AB) = \int_A^B n\sin\phi ds \qquad (5.38)$$

where ϕ is the angle of incidence of a ray on the line element ds. Thus,

$$I_\alpha(AB) = \langle n\sin\phi \rangle L_{AB} \qquad (5.39)$$

where < > denotes the average and LAB is the length of the curve from A to B.

It follows that, provided the aperture AB is filled with rays from the line of point sources just indicated, the étendue is simply

$$I_\alpha(AB) - I_\beta(AB)$$

But

$$I_\alpha(AB) = [P_\alpha B] - [P_\alpha A] \qquad (5.40)$$

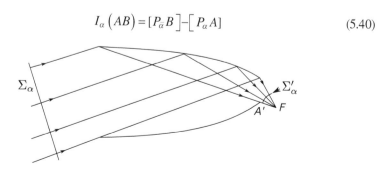

FIGURE 5.28   The edge-ray principle as a limiting case of matching wave fronts.

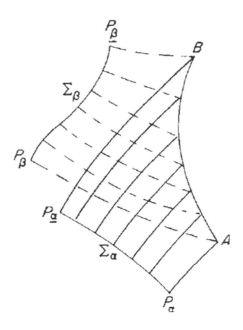

FIGURE 5.29 Calculating the étendue.

where the square brackets denote optical path lengths along the rays so that we obtain, for the étendue,

$$H = \left[P_{\bar{\alpha}} B \right] + \left[P_{\beta} A \right] - \left[P_{\alpha} A \right] - \left[P_{\bar{\beta}} B \right] \tag{5.41}$$

This result can be seen to be a simple generalization of the result of Equation (5.11).

Now let there be some kind of system constructed that achieves the desired transformation of incident extreme pencils with emergent extreme pencils, as in Figure 5.30. The system takes Σ_α into $\Sigma_{\alpha'}$ and Σ_β into $\Sigma_{\beta'}$, and we want it to do so without loss of étendue. We write down the optical path length from $P_{\bar{\alpha}}$ to $P_{\bar{\alpha}}'$ and equate it to that from P_α to P_α'. Similarly, for the other pencil,

$$\left[P_{\bar{\alpha}} B \right] + \left[BA' \right]_{\bar{\alpha}} + \left[A'P_{\bar{\alpha}}' \right] = \left[P_\alpha A \right] + \left[AB' \right]_\alpha + \left[B'P_\alpha' \right]$$

$$\left[P_\beta A \right] + \left[AB' \right]_\beta + \left[B'P_\beta' \right] = \left[P_{\bar{\beta}} B \right] + \left[BA' \right]_{\bar{\beta}} + \left[A'P_{\bar{\beta}} \right] \tag{5.42}$$

where

$$\left\{ \left(\left[P_\beta A \right] - \left[P_\alpha A \right] \right) - \left(\left[P_{\bar{\beta}} B \right] - \left[P_{\bar{\alpha}} B \right] \right) \right\}$$

$$- \left\{ \left(\left[A'P_{\bar{\beta}}' \right] - \left[A'P_\alpha' \right] \right) - \left(\left[B'P_\beta' \right] - \left[B'P_\alpha' \right] \right) \right\} \tag{5.43}$$

$$= \left[AB' \right]_\alpha - \left[AB' \right]_\beta + \left[BA' \right]_{\bar{\beta}} - \left[BA' \right]_{\bar{\alpha}}$$

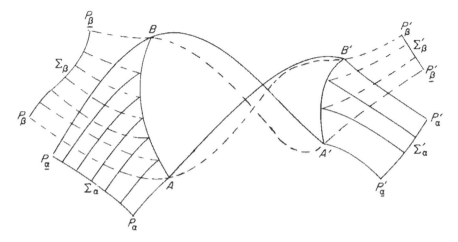

FIGURE 5.30   Rays inside the concentrator.

The left-hand side of this equation can be as the difference between the étendues at the entry and exit apertures. Since we require this difference to vanish, we have to make the right-hand side of the equation equal to zero vanish. A simple way to do this would be to ensure that the optical system of the concentrator is such that the α and β ray paths from A to B′ coincide, and the same for those from B to A′. We can do this by starting segments of mirror surfaces at A and A′ in such directions so as to bisect the angles between the incoming α and β rays. In this way we effectively make the $P_\alpha A$ and $P_\beta A$ the same through the optical system between AB and A′B′. In other words, the two extreme rays become one. We then continue the mirror surfaces in such a way as to make all β rays join up with the corresponding emerging β' rays; in other words, we image the β pencil exactly into the β' pencil, and similarly for the other mirror surface connecting B and B′. We have thus completed the construction and used up all degrees of freedom in doing so. This is just one of all the many ways to satisfy Eq. (5.43). Simultaneous Multiple Surfaces (SMS) has been able to also achieve similar results.

5.15   A CONSTRUCTIVE DESIGN PRINCIPLE
FOR OPTIMAL CONCENTRATORS

We conclude this chapter with a discussion of a design prescription for optimal concentrators that historically has been a fertile source of new and useful solutions. In contrast to the edge-ray principle, which is allied to such abstract notions as étendue and the Hilbert integral theorem, this practical procedure directly instructs us how to draw the profile curve of the mirror. The statement is simply that we maximize the slope of the mirror profile curve consistent with reflecting the extreme entrance rays onto the absorber and subject to various subsidiary conditions that we may wish to impose. This generic principle for designing optimal concentrators is inherent in the earliest references (Hinterberger & Winston, 1966a,b).

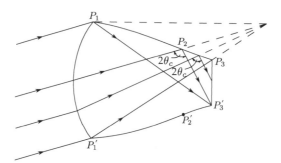

FIGURE 5.31 The solid concentrator that just satisfies the critical angle condition. The profile is hyperbolic from P_1 to P_2 (assuming the entry surface is aberration-free). Then if the critical angle θ_c for total internal reflection is reached at P_2, it is an equiangular spiral.

To see how this rule operates we notice that the CPC designs for variously shaped absorbers are obtained when we don't impose any subsidiary conditions. On the other hand, the θ_i/θ_o design results when we impose the condition that the maximum exit angle not exceed θ_o. This approach is particularly useful in situations where the subsidiary conditions preclude an ideal system—hence strict application of the edge-ray principle. For example, the various totally internally reflecting designs are obtained by specifying total internal reflection at the external wall as a condition. As discussed earlier in this chapter, when the input ray distribution and index of refraction fall outside a specified range, ideal solutions are not possible. However, efficient designs are still available.

Consider the example in Figure 5.31, where the front entrance face is curved and the index of refraction would not be sufficiently high to totally reflect the extreme rays along the portion P_2, P_3, were we to follow the edge-ray principle. The maximum slope rule would suggest we maintain a constant angle (the critical angle) between the extreme ray and normal to the wall along this portion of the profile curve. In the approximation that the curve face images the extreme pencil onto a point, the portion between P_1, P_2 is a hyperbola, whereas the portion between P_2, P_3 is an arc of an equiangular spiral.

REFERENCES

Bassett, I. M., and Derrick, G. H. (1978). The collection of diffuse light onto an extended absorber. *Opt. Quantum Electron.* **10**, 61–82.

Hinterberger, H., and Winston, R. (1966a). Efficient light coupler for threshold Cerenkov counters. *Rev. Sci. Instrum.* **37**, 1094–1095.

Hinterberger, H., and Winston, R. (1966b). Gas Cerenkov counter with optimized light-collecting efficiency. *Proc. Int. Conf. Instrum. High Energy Phys.* 205–206.

Hottel, H. (1954). Radiant heat transmission. In *Heat Transmission*, 3rd Ed. (W. H. McAdams, ed.) McGraw-Hill, New York.

Luneburg, R. K. (1964). *Mathematical Theory of Optics.* University of California Press, Berkeley, CA. This material was originally published in 1944 as loose sheets of mimeographed notes and the book is a word-for-word transcription.

Ortobasi, U. (1974). Proposal to Develop an Evacuated Tubular Solar Collector Utilizing a Heat Pipe. Proposal to National Science Foundation (unpublished). Private communication to the author.

Ploke, M. (1967). Lichtführungseinrichtungen mit starker Konzentrationswirkung. *Optik* **25**, 31–43.

Rabl, A., Goodman, N. B., & Winston, R. (1979). Practical design considerations for CPC solar collectors. *Solar Energy* **22**(4), 373–381.

Winston, R. (1974). Principles of solar concentrators of a novel design. *Sol. Energy* **16**, 89–95.

Winston, R. (1976). Dielectric compound parabolic concentrators. *Appl. Opt.* **15**, 291–292.

Winston, R. (1978). Cone collectors for finite sources. *Appl. Opt.* **17**, 688–689.

Winston, R., and Hinterberger, H. (1975). Principles of cylindrical concentrators for solar energy. *Sol. Energy* **17**, 255–258.

6 The Flowline Method for Nonimaging Optical Designs

6.1 THE CONCEPT OF THE FLOWLINE

We have seen that the edge-ray approach to nonimaging concentration has considerable flexibility in handling 2D problems. Its application in 3D is more limited, and, although useful designs are obtainable in 3D, these fall short of ideal concentration. This motivates the search for alternatives to edge-ray designs, especially in 3D applications (Winston & Welford, 1979a).

In the geometrical optics approximation, we describe the propagation of light rays as trajectories in a six-dimensional phase space (p, x), where the components of x are the generalized coordinates (x_1, x_2, x_3) and the components of p are the generalized momenta. These momenta give the optical path length when the light ray traverses length ds through a medium with an index of refraction n by the relation $nds = p_1 dx_1 + p_2 dx_2 + p_3 dx_3$. For example, in cartesian coordinates (x, y, z) the components of p are just the optical-direction cosines.

It is useful to picture the propagation of the totality of light rays as if it were a fluid flow in this six-dimensional space. This flow is subject to conservation theorems that are essentially geometric and follow from the law of propagation of light rays. These are loosely referred to as Liouville's theorem, but they are more properly called the integral invariants of Poincaré. For our purpose, the most useful invariant is constructed out of a vector $\vec{J}$, whose components are

$$\vec{J}_1 = \int dp_2 dp_3, \vec{J}_2 = \int dp_1 dp_3, \vec{J}_3 = \int dp_1 dp_2 \qquad (6.1)$$

Postulating as always a medium with no attenuation or radiation sources, the surface integral of $\vec{J}$, $\int \vec{J} \cdot d\vec{A}$ is invariant so that $\vec{J}$ has zero divergence. We shall call $\vec{J}$ the geometrical vector flux and the direction of $\vec{J}$ the flowline. Quantities like $\vec{J}$ occur in radiative transfer where $\vec{J} \cdot d\vec{A}$ is proportional to the net flux through the surface element dA and in photometry where $\vec{J}_z$ is proportional to the illumination on the x, y plane. This is most readily seen by rewriting

$$J_z = \int n^2 d\Omega \cos\theta_z \qquad (6.2)$$

where θ_z, is the inclination of the elementary beam to the z axis.

6.2 LINES OF FLOW FROM LAMBERTIAN RADIATORS: 2D EXAMPLES

All Lambertian radiators that subtend the same solid angle at a point, in the strict sense that they are bounded by the same cone, give the same vector flux $\vec{J}$ at that point. This can be made clear by some simple examples (see Winston & Welford, 1979b).

Consider a strip AA' in the x, y plane, of width 2 in the y direction, as in Figure 6.1. A simple calculation shows that $\vec{J}$ points along the bisector of the angle APA', and its magnitude is $\vec{J} = 2\sin\theta$, where the angle $APA' = 2\theta$. From elementary geometry the lines of flow are confocal hyperbolae with A and A' as foci. Notice that the difference of distances from P to the foci $|AP| - |A'P|$ is constant along a flowline from the property of hyperbolae. As a second example, Figure 6.2 shows a semi-infinite strip. It is easily seen that the flowlines are confocal parabolas, with the focus at the end A of the strip. A final example is a wedge: Figure 6.3 shows an example where the wedge angle is 60°. In regions where only one face of the wedge is seen, the pattern is the same as for the semi-infinite strip of Figure 6.2. From any point where both faces are seen, the effect is of a uniform Lambertian source subtending an angle of 60°. Thus $|\vec{J}|$ is constant in this region, and the lines of flow are parallel as indicated.

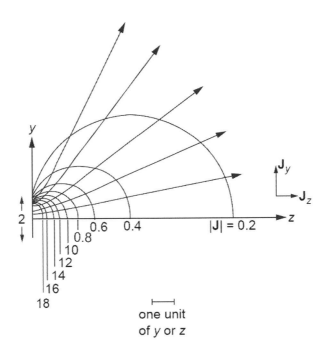

FIGURE 6.1 Lines of flow and loci of constant $|\vec{j}|$ for the two-dimensional strip: the width is taken as two units. The axes of circles are labeled with the values of $|\vec{j}|$. The two sets of lines are not orthogonal because $\vec{J}$ is not derived from a scalar potential.

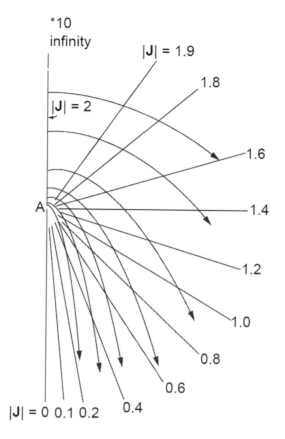

FIGURE 6.2 Lines of flow and loci of constant $\left|\vec{J}\right|$ for a semi-infinite strip.

Figure 6.4 shows a truncated 60° wedge $RQQ'R'$, which is infinitely extended back from the apex with an included angle $2\theta=60°$. We have to consider four different regions in calculating vector flux fields; these are separated by the boundary lines of the wedge produced, as shown in broken lines, and they are labeled regions A (and A'), B, C (and C'), and D. A point P in region A receives flux only from the face RQ so that the $\left|\vec{J}\right|$ distribution is the same as for a semi-infinite strip; that is, the flowlines are confocal parabolas with focus at Q, and the loci of constant $\left|\vec{J}\right|$ are straight lines radiating from Q. If P is in region B, it receives flux only from the strip QQ', and again we have the result that the lines of flow are confocal hyperbolas with Q and Q' as foci, and the loci of constant $\left|\vec{J}\right|$ are arcs of circles through QQ and Q'. If P is anywhere in region D, the source has the constant angular subtense 2θ; thus $\left|\vec{J}\right|$ is constant at $2\sin\theta$ throughout this region, and the lines of flow are straight lines perpendicular to QQ'. Finally, in region C, point P receives radiation from RQ and QQ', and the angular subtend is 2θ as shown in Figure 6.14. The direction of $\vec{J}$ thus bisects the angle between a constant direction (parallel to QR) and the radius from a fixed point Q'; thus, the lines of flow in region C are confocal parabolas with focus at

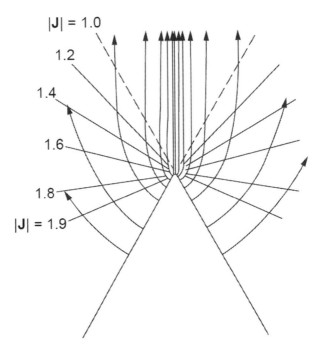

FIGURE 6.3   Lines of flow and loci of constant $\left|\vec{j}\right|$ for a wedge.

Q' and axis through Q' and parallel to RQ. The loci of constant $\left|\vec{J}\right|$ are straight lines radiating from Q', since the angle 2θ as constructed is constant along these lines. The lines of flow for a truncated wedge are shown in Figure 6.5. Recent advancement of flow line has shown that the flow lines within the ideal nonimaging concentrators the also form ideal nonimaging concentrators (Jiang 2016a, 2016b).

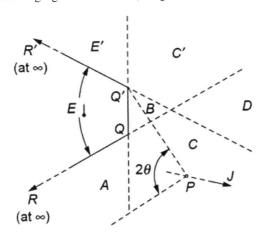

FIGURE 6.4   A truncated wedge of angle 2θ. The truncated QQ' is shown as forming equal angles with the two sides of the wedge, but in fact the argument is still valid if this is not so.

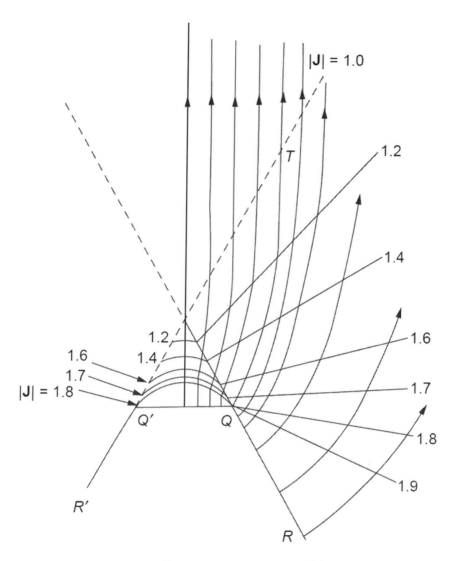

FIGURE 6.5 Lines of flow of $\vec{J}$ and labeled loci of constant $\left|\vec{j}\right|$ for the truncated wedge; only the loci below the line of symmetry are shown. The wedge angle is 60°.

6.3 3D EXAMPLE

A disk as a Lambertian radiator—that is, the circular opening of a blackbody cavity—is the most important example to consider. The components of $\vec{J}$ can be calculated by integrating expressions such as Equation (1.1) over the surface of the disk.

These integrations have been carried out for radiative transfer purposes, and we shall simply quote the results. Sparrow and Cess (1978), for example, provide the details, and they also give a useful transformation of the surface integral over the disk into a line integral around its edge; this considerably simplifies the calculation.

Let c be the radius of the disk and take the coordinate axes as in Figure 6.6.

Then the components of $\vec{J}$ in the y, z plane, which from symmetry are all that need be considered, are given by

$$\vec{J}_y = \frac{\pi}{2}\frac{z}{y}\left\{\frac{c^2 + y^2 + z^2}{\left[\left(c^2 + y^2 + z^2\right)^2 - 4c^2 y^2\right]^{\frac{1}{2}}} - 1\right\}$$

$$\vec{J}_z = \frac{\pi}{2}\left\{\frac{c^2 - y^2 - z^2}{\left[\left(c^2 + y^2 + z^2\right)^2 - 4c^2 y^2\right]^{\frac{1}{2}}} + 1\right\} \qquad (6.3)$$

The slope of the lines of flow is given by arctan (J_y/J_z) and when this is calculated, it turns out that the lines of flow are the same hyperbolas as for the 2D case. The lines of flow in three dimensions are obtained by rotation about the z axis, as shown in Figure 6.7.

FIGURE 6.6 Geometry for calculating $\left|\vec{J}\right|$ for a luminous disk. Notice here x represents the radial direction instead of the Cartesian coordinates.

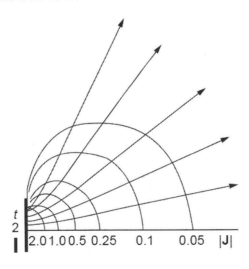

FIGURE 6.7 Lines of flow and loci of constant $\left|\vec{J}\right|$ for a luminous disk.

6.4 A SIMPLIFIED METHOD FOR CALCULATING LINES OF FLOW

In certain cases it is easier to calculate the étendue than to perform the preceding integrals. In cases where the étendue is known, we can use the solenoidal property of $\vec{J}$ to find the lines of flow. For example, in the case described by Figure 6.8, the étendue H from F, F' to A, A' is equal to $\left[A'F\right]-\left[A'F'\right]$ in the 2D case and $\pi^2\left\{\left[A'F\right]-\left[A'F'\right]\right\}^2/4$ in the 3D case (the Hottel strings!) Then holding $H=$constant means we intercept a constant flux of $\vec{J}$ and therefore move along a line of flow. But holding the difference of distances to two foci, F, F' constant is the geometric definition of a hyperbola. Hence we recover the hyperbolic flowlines with lighter computation.

Assuming that there is a vector potential for $\vec{J}$, $\vec{\nabla}\times\vec{A}=\vec{J}$, and due to the axisymmetric configuration of the problem, we can assume that only the x direction has nonzero value A_x. Therefore the étendue H can also be represented by $H=2\pi A_x$. Because $J_z=\left(1/y\right)\left(\partial\left(yA_x\right)/\partial y\right)$, $J_y=-\partial A_x/\partial z$, we can also obtain the components of $\vec{J}$ due to its being the curl of $\vec{A}$. Thus, the relations $J_z=\dfrac{1}{2\pi y}\left(\dfrac{\partial H}{\partial y}\right)$ and $J_y=-\dfrac{1}{2\pi y}\left(\dfrac{\partial H}{\partial z}\right)$ give Equation (6.3).

6.5 PROPERTIES OF THE LINES OF FLOW

Let $\delta\Sigma$ be a small surface element with a 100% reflecting mirror surface on one side. If we place $\delta\Sigma$ in a vector flux distribution at a point P (Figure 6.9) so the lines of flow lie in its surface, then the lines of flow locally on the mirror side will be undisturbed. To see this, we note that before the mirror is placed at position P, the resolved part of $\vec{J}$ normal to the surface is zero, since $\vec{J}$ lies along the lines of flow. Now this zero normal component is made up of equal and opposite components $\vec{J}_n^+$ and $\vec{J}_n^-=-\vec{J}_n^+$ from rays coming from either side of $\delta\Sigma$. But the effect of inserting the

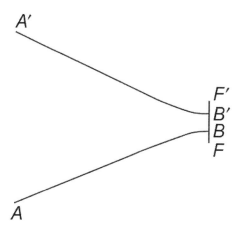

FIGURE 6.8 Étendue for a luminous disk.

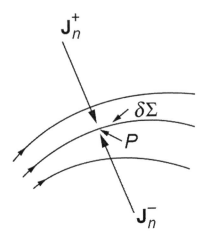

FIGURE 6.9 Placing a mirror in a vector flux field.

mirror is to remove $\vec{J}_n^-$ and replace it by $\vec{J}_n^+$ with reversed sign, according to the law of reflection, so there is no local disturbance.

The insertion of a mirror can, of course, disturb the lines of flow at some distance because it does not follow from $\vec{J}_n^- = -\vec{J}_n^+$ that each ray from one side is paired with a ray in the same plane of incidence and with the same angle of incidence on the other side. However, this simple situation does hold for a 2D vector flux field illuminated by a Lambertian radiator without obstructions and in the important case of a 3D field produced by a Lambertian disk.

6.6 APPLICATION TO CONCENTRATOR DESIGN

We have noted that a mirror surface placed along lines of flow of the vector flux will not disturb the overall flow pattern under a certain condition. This condition is that the zero normal component of $\vec{J}$ must result from the fact that, for each ray r_1 incident on one side of the surface, there must be a corresponding ray r_2 incident on the other side such that $r2$ is in the direction of the reflection of r_1; we call this a condition of detailed balance of the rays. It is possible to have $\vec{J}_{\text{normal}} = 0$ at a surface containing the lines of flow without the condition of detailed balance being fulfilled, and then $\vec{J}$ would remain undisturbed locally but not over the entire illuminated region. Clearly, the condition of detailed balance holds for a 2D field illuminated by a Lambertian source without obstructions.

We may therefore place mirrors along any of the flowlines of Figure 6.5 without disturbing the flux patterns. For example, suppose we place a mirror along the flowline from Q, ending at the point T where the flowline meets the continuation of the wedge boundary $R'Q'$, and similarly another mirror $Q'T'$ symmetrically on the other side. These mirrors would take the Lambertian flux $\left(\left|\vec{j}\right| = 2\right)$ from QQ' and convert it to a uniform flux with $\left|\vec{j}\right| = 1$ at TT'. The rays in the emergent beam must spread over an angle $\pm 30°$, since $\left|\vec{j}\right| = 1$. Clearly, we have constructed a concentrator in reverse, and our construction with parabolas with tilted axes is precisely that for the compound parabolic concentrator.

Since light rays, and therefore flowlines of $\vec{J}$, are reversible, we see that a way to construct a concentrator with maximum theoretical concentration is to place mirrors in the flowlines under conditions of detailed balance of the rays.

This constitutes an entirely new perspective on concentrator design.

6.7   THE HYPERBOLOID OF REVOLUTION AS A CONCENTRATOR

The results of the previous section lead to another interesting new perspective on the design of nonimaging concentrators. Suppose we start with the disk radiator and we place a mirror in the form of a hyperboloid of revolution coincident with a set of flowlines as in Figure 6.10. We truncate the mirror at some distance so the open end is a circle of diameter AA'. Considering the inside as a mirror, this forms a nonimaging concentrator with unusual properties. The foci of the hyperbolas in the section shown are at F and F', the ends of the diameter of the original disk.

Then all rays entering the aperture AA' and pointing somewhere inside the disk will be reflected by the mirror so as to strike, eventually, the inner disk of diameter BB'. Thus, the concentrator takes all rays from the virtual source FF', which can pass the entry aperture AA' and concentrates them into an exit aperture BB'.

This result is easily proved for rays in a meridional section—that is, in the plane of the diagram. Consider a ray passing through A' and aimed at F, as indicated by the double arrow; after the first reflection, it will, by a basic property of conic sections, be aimed at F', and so on as indicated. Thus, the extreme angle rays emerge from the

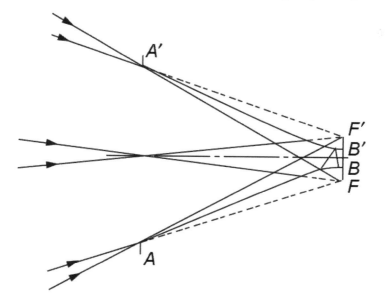

FIGURE 6.10   The hyperboloid of revolution as a concentrator.

exit aperture but only after an infinite number of reflections. It is easily seen that rays at angles inside the extreme angles all emerge. Thus, in the meridional plane this is a concentrator of maximum theoretical concentration.

This property also holds for skew rays, although this is not quite so obvious. We give proof in Appendix F that any skew ray incident on the mirror and pointing on the rim of the disk FF' will be reflected to another point on the rim and the result follows directly. When used in reverse, the same design produces a virtual ring that fills the space between a Lambertian source of diameter BB' and the larger diameter FF'. The visual effect produced is striking.

6.8 ELABORATIONS OF THE HYPERBOLOID: THE TRUNCATED HYPERBOLOID

The properties of the truncated hyperboloid are evident from the preceding discussion.

We refer to Figure 6.11, where the hyperboloid stops short of the plane of the virtual source. For example, rays collected from the source of diameter FF' by the small aperture will appear to originate from the same source upon emerging from the large aperture. Alternatively, rays directed toward the source of diameter FF' at the large aperture will illuminate the same source upon exiting the small aperture. The truncated hyperboloid may be useful on its own, but usually it is in combination with lenses, as discussed in the following.

6.9 THE HYPERBOLOID COMBINED WITH A LENS

We have seen that, for the bare hyperboloid, both real and virtual Lambertian sources are in the same plane. This is a limitation in certain applications; however, we can manipulate these positions with lenses to place the sources at more convenient locations. We show in Figure 6.12 the full hyperboloid with a positive lens of diameter $AA'=2a$ at the large aperture, which takes the source plane to infinity—that is, the source is in the focal plane of the lens. Consider the properties of this design. For simplicity we neglect off-axis aberrations of the lens; a complete discussion is found in Welford, O'Gallagher, and Winston (1987). We only remark that, since the requirements on the lens need not be severe (e.g., the focal ratio may be chosen large, as will be shown), the aberrations can be made negligible in practice. In this approximation, the cone of rays of angle θ is imaged by the lens onto the circle of diameter $FF'=2b$. The hyperboloid collects the rays that fill this cone and concentrates them into a smaller circle of diameter $BB'=2c$. We show in Appendix G that $c/a=\sin\theta$. Therefore, this is an ideal concentrator. This result could have been anticipated from

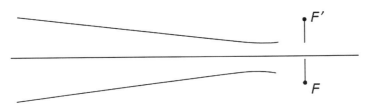

FIGURE 6.11   The truncated hyperboloid of revolution as a concentrator.

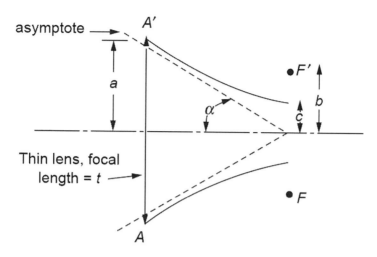

FIGURE 6.12 The lens/flow-line concentrator. The entry aperture is FF', and the exit aperture is at the waist of the hyperboloid. The semiangle of the asymptotic cone is c/b, and $\left(\dfrac{a}{c}\right)^2$ is the concentration ratio.

the fact that we have postulated an ideal lens while the hyperboloid on its own is an ideal concentrator, albeit operating on a virtual source.

The design considerations readily follow from the geometry of a hyperbola.

With reference to Figure 6.12, the focal ratio of the lens can be estimated from the slope α of the asymptote to the hyperbola. The slope is given by $\sin \alpha = c/b$; the acceptance angle θ determines the focal length of the lens, completing the design.

6.10   THE HYPERBOLOID COMBINED WITH TWO LENSES

We have seen that we can construct a 3D concentrator approaching ideal properties by placing a suitable lens at the large aperture of the hyperboloid. Similarly, we can construct a 3D θ_1/θ_2 concentrator (angle transformer) by placing an additional negative lens at the small aperture of the truncated hyperboloid, which moves the source plane to infinity as shown by Ning (1988). The construction is shown in Figure 6.13, where it is apparent that the focal lengths of the two lenses are related by $\dfrac{F_1}{F_2} = \dfrac{\tan \theta_1}{\tan \theta_2}$, whereas their focal ratios are related as $\dfrac{f_1}{f_2} = \dfrac{\cos \theta_1}{\cos \theta_2}$. We see that this design lends itself to all dielectric versions, since the positive and negative lenses are readily molded into the shape of the transformer, whereas in practical cases the condition for total internal reflection is likely to be satisfied.

6.11   GENERALIZED FLOWLINE CONCENTRATORS
WITH REFRACTIVE COMPONENTS

In 2D it is possible to generalize the flowline construction following the notions expressed in Section 6.4. For simplicity, we confine our discussion to systems that are symmetrical transverse to the optic axis. As already noted in Section 6.4, the

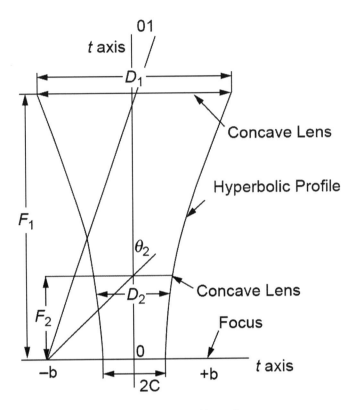

FIGURE 6.13 θ_1/θ_2 transformer using an FLC with two lenses.

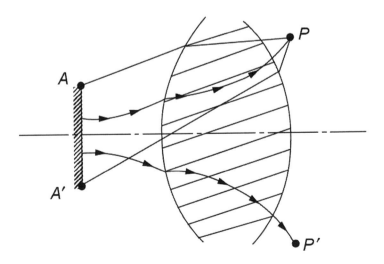

FIGURE 6.14   Flowlines with refractive components AA' are a Lambertian source. The arrows indicate row lines; the plain lines, rays.

étendue H generalizes to the difference of optical path lengths (up to an overall constant). This remains true even in the presence of refractive media, provided the optical path lengths are measured along rays. These rays need not be straight lines. Thus, in Figure 6.14, the étendue H from Lambertian source AA' to section PP' is proportional to $[A'P] - [AP]$, where the brackets indicate optical path lengths. It follows that the lines of flow (indicated by arrows in the figure) lie along contours of H=constant. Since the detailed balance condition holds in 2D, we may construct concentrators by placing mirrors along the flowlines. However, it does not follow that the 3D construction obtained by rotating the 2D flowline about the optic axis will automatically satisfy detailed balance. Specific cases will have to be checked with respect to detailed balance before the usefulness of the 3D designs can be evaluated.

REFERENCES

Jiang, L., and Winston, R. (2016a). Flow line asymmetric nonimaging concentrating optics. In *Nonimaging Optics: Efficient Design for Illumination and Solar Concentration XIII—Commemorating the 50th Anniversary of Nonimaging Optics* (Vol. 9955, p. 99550I). International Society for Optics and Photonics.

Jiang, L., and Winston, R. (2016b). Thermodynamic origin of nonimaging optics. *Journal of Photonics for Energy,* **6**(4), 047003.

Ning, Xiaohui. (1988). Three-dimensional ideal θ_1/θ_2 angular transformer and its uses in fiber optics. *Appl. Opt.* **27**(19), 4126–4130.

O'Gallagher, Joseph, Winston, Roland, and Welford, Walter T. (1987). Axially symmetric nonimaging flux concentrators with the maximum theoretical concentration ratio. *JOSA A* **4**(1), 66–68.

Sparrow, E. M., and Cess, R. D. (1978). *Radiation Heat Transfer.* Hemisphere, Washington, DC.

Winston, R., and Welford, W. T. (1979a). Geometrical vector flux and some new imaging concentrators. *J. Opt. Soc. Am. A* **69**, 532–536.

Winston, R., and Welford, W. T. (1979b). Ideal flux concentrators as shapes that do not disturb the geometrical vector flux field: A new derivation of the compound parabolic concentrator. *JOSA* **69**(4), 536–539.

7 Freeform Optics and Supporting Quadric Method

Introduction

This chapter along with Chapters 8 and 9 are devoted to optical design with freeform reflecting surfaces. We begin by summarizing some of the notations and terminology used in these chapters.

NOTATION

$\mathbb{R}^2$, $\mathbb{R}^3$:	The two- and three-dimensional vector spaces, respectively.
$\mathbb{S}^2$:	The sphere of unit radius with center at the origin O of a Cartesian coordinate system in $\mathbb{R}^3$.
A:	An arbitrary subset of $\mathbb{R}^2$.
$C^k(A)$:	The set of k times continuously differentiable functions defined on a set A; k is a nonnegative integer.
$L(A)$:	The set of functions on a set A such that $\int_A |f(x)|\,dx < \infty$.
$\|f\|_{L(A)} := \int_A |f(x)|\,dx$:	The norm of f on $L(A)$.
$C^k(A) \cap C^l(B)$:	Functions which are in class $C^k(A)$ and in $C^l(B)$.
$C^1(A) \times C^1(B)$:	Set of pairs of functions f,g such that $f \in C^1(A)$, $g \in C^1(B)$, $A, B \subset \mathbb{R}^2$.
Radiance:	($\equiv$ radiance pattern) I; radiant flux per unit source area, W/m^2.
Irradiance:	($\equiv$ irradiance pattern) L; radiant flux per unit target area, W/m^2.

For a point source the input aperture is always taken on the unit sphere $\mathbb{S}^2$ with center at the source. The source radiance ($\equiv$ intensity or radiance intensity) in this case is W/sr.

In the far-field problem the reflected *directions* are identified with points on the unit sphere $\mathbb{S}^2$. The target irradiance (often referred to as intensity or irradiance intensity) in this case is the flux per unit area of the aperture on $\mathbb{S}^2$ representing the directions of reflected rays, W/sr; see Figure 7.1.

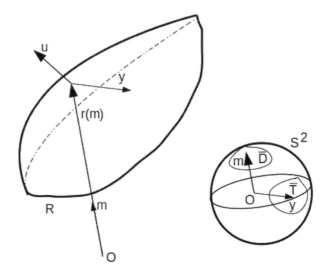

FIGURE 7.1   Far-field reflector design problem.

The optical systems considered in this book are required to redirect the light and reshape the radiance pattern of a source so that the total output irradiance is distributed on a given target and has a pattern prescribed in advance. Following Winston et al. (2005) we refer to problems of this kind as prescribed irradiance problems. The goals of the design, in addition to the redirection and reshaping of the flow of light, may also include efficiency, compactness of the device, and optimized performance according to some criteria. Consider, for example, the problem of designing a reflector system shown schematically in Figure 7.1.

Here and in the rest of the book the framework for formulating prescribed irradiance problems is **geometrical optics** (GO). Afterwards, the designed optics can be tested computationally and/or experimentally by other techniques, including physical optics.

Let O denote a nonisotropic point source of light placed at the origin (also denoted by O) of a Cartesian coordinate system in 3D-space $\mathbb{R}^3$. The source O emits light rays in a set of directions defined by the aperture $\bar{D}$ $(= D \cup \partial D$, where ∂D is the boundary of D) given as a closed set on a sphere $\mathbb{S}^2$ of unit radius centered at O. Points on $\mathbb{S}^2$ are treated also as unit vectors in $\mathbb{R}^3$ with initial points at O. Usually, we assume that $\bar{D}$ is a convex domain on $\mathbb{S}^2$.

We denote by $I(m)$ the radiance of the source O in direction $m \in D$ and assume that I is a nonnegative and integrable function on D. Suppose that a light ray emitted by the source O in direction $m \in \bar{D}$ is incident on a smooth perfectly reflecting surface R at point $r(m)$ with unit normal $u = u(m)$ and reflects off it in direction $y = y(m)$, $|y(m)| = 1$. In this case, the aim is to design a system with one reflecting surface R (to be determined!) intercepting the light beam from the source O and redistributing the light with prescribed irradiance over a given region T in the far field. The input and output irradiances are not shown in Figure 7.1. This example illustrates the type of design problems which can be analyzed by methods described in Chapters 8 and 9.

Many applications require optical systems capable of controling output irradiance. Such systems are needed in nonimaging optics (concerned with light concentration, illumination, and imaging applications), laser optics (concerned with reshaping laser beams), and the design of reflector antennas with prescribed directivity properties, to name just a few (Winston et al., 2005, Schruben, 1974; Malyak, 1992; Burkhard & Shealy, 1987; Oliker et al., 1994; Kinber, 1962; Kinber, 1984; Galindo-Israel et al., 1979; Galindo-Israel et al., 1987; Galindo-Israel et al., 1991a; Galindo-Israel et al., 1991b; Westcott, 1983; Westcott & Norris, 1975; Oliker, 2006a; Rhodes & Shealy, 1980; Dickey, 2014; Dickey et al., 2005).

Traditional methods for designing optics for prescribed irradiance problems are often based on an a priori assumption that the sources, targets, and light patterns have rotational or some other symmetry. However, designers of optics often encounter practical tasks in which optical surfaces are expected to accommodate light sources, targets, and illumination patterns without special symmetries such as rotational or rectangular. In these cases, traditional designs constrained by an a priori set-particular symmetry may lead to highly energy inefficient devices.

In contrast, **freeform** lenses and mirrors are expected to be designed without any a priori assumed symmetry and therefore significantly more degrees of freedom are available to the designer to create optical systems with desired characteristics (Winston et al., 2005; Ries & Muschaweck, 2002; Berry, 2017; Oliker, 2017; Oliker, 2003; Oliker, 2005a; Oliker, 2007; Oliker, 2005b; Oliker, 2006b). A freeform mirror shown in Figure 7.2 provides an example of the complexity of a shape representable by a freeform optical surface.

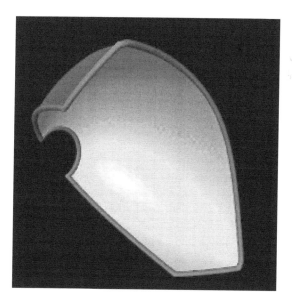

FIGURE 7.2   Freeform reflector transforming a sinusoidal shaped radiation pattern from a point source into a uniform planar far-field irradiance; see Section 9.4 for more details. Designed by V. Oliker using supporting quadric method; I/O data provided by J. C. Miñano and P. Benítez (Benítez et al., 2006).

Unfortunately, in tasks involving freeform surfaces, traditional design methods in most cases can not be used and the development of new design methods capable of utilizing freeform surfaces is necessary. This area of research and development is presently at an early stage.

Theoretical and computational methods for designing freeform mirrors for applications which require the generation of prescribed irradiances is the main subject of this work. Freeform lenses will be discussed in a separate publication. For mirrors specifically we present here rigorous approaches to optical design producing effective tools for development, analysis, and description of required freeform optics. Such description may be a collection of formulas or functions defining the designed optical surfaces or an algorithm clearly indicating the steps leading to quantitative definitions of required surface(s).

We outline now the issues involved in the development of methods for design with freeform surfaces. In the GO approximation the derivation of equations of freeform optical surface(s) transforming a given input radiation into a prescribed irradiance pattern is based on a systematic application of the ray-tracing equations and energy conservation law. Physically, the resulting equations express the fact that the radiance of the source and the required output irradiance are related by a factor which is the determinant of the Jacobian of the ray-tracing map (defined by the reflecting/refracting surface[s] to be determined!) The required optical surfaces are determined by solving these equations. Depending on the approach to derivation of the Jacobian equation, this step requires careful analysis of the theoretical and numerical solvability of either a nonlinear system of first order partial differential equations (PDEs) (Galindo-Israel et al., 1979; Galindo-Israel et al., 1987; Galindo-Israel et al., 1991a; Galindo-Israel et al., 1991b) or a nonlinear second order PDE of Monge-Ampère type for the function(s) describing the desired optical surface(s) (Winston et al., 2005; Kinber, 1962; Westcott, 1983; Westcott & Norris, 1975; Bösel & Gross, 2017; Oliker, 2014; Oliker, 2003). The resulting equations are quite complex even when it is assumed that the source is small or a collimator.

Solving numerically such equations is a very challenging problem. Because of strong nonlinearities the traditional discretization schemes based on finite differences in general fail to converge (Oliker & Prussner, 1988) and, for the same reason, the discretization schemes based on projection methods are not applicable. This is true for problems formulated as systems of PDEs (Galindo-Israel et al., 1987; Galindo-Israel et al., 1991a; Galindo-Israel et al., 1991b) and as Monge-Ampère equations (Westcott, 1983; Oliker, 2003). The engineering literature contains numerous papers in which attempts are made to solve the equations of beam-shaping numerically with ad hoc approaches. However, the authors almost always rely on specific numerical examples to justify their approaches and results (Galindo-Israel et al., 1979; Galindo-Israel et al., 1987; Galindo-Israel et al., 1991a; Galindo-Israel et al., 1991b; Westcott, 1983). This is a serious drawback making the reliability of such approaches problematic.

Alternative formulations of some of such design problems based on variational considerations are sometimes possible. For example, it has been shown that, when the desired irradiance pattern is in the far field or when two refracting/reflecting surfaces are used, the design problems can be formulated as problems of linear programming (Glimm & Oliker, 2003; Glimm & Oliker, 2004; Oliker, 2011). In the near-field case for a single lens, however, this formulation is impossible (Graf & Oliker, 2012).

In recent years a very general geometric framework for solving several classes of equations arising in beam-shaping problems has been developed and applied by the author of this survey and his collaborators in Oliker et al. (1994); Oliker and Prussner (1994); Kochengin and Oliker (1998); Oliker (1989); Newman and Oliker (1994); Kochengin and Oliker (1997); Kochengin et al. (1998); Caffarelli and Oliker (2008); Caffarelli et al. (1999); and Kochengin and Oliker (2003). The developed "Supporting Quadric Method" (SQM) is based on new geometric ideas for constructing the so-called "weak" solutions of Monge-Ampère equations. It provides a rigorous and unified framework for establishing existence of solutions and calculating them numerically. However, weak solutions representing sag functions, while continuous, may not be differentiable at all points and the resulting optical surface may have straight and curved edges and corner points. It is important to note that the set of all points at which nondifferentiability occurs has zero area and with some additional efforts weak solutions can be smoothed out.

Regarding numerical calculations required for explicit representation of the sag functions of the designed optics, the SQM, in essence, includes a built-in numerical procedure. In particular, for problems with one reflecting surface, specific computational algorithms based on SQM were proposed in Kochengin and Oliker (1998) and Caffarelli et al. (1999). The algorithms described in these papers are motivated by the physical and geometric properties of weak solutions to these problems (see Section 2.9 for more details) and are proven to converge. However, the rate of convergence of the corresponding numerical algorithms may become low when the number of discretization nodes is high. Such behavior is typical for strongly nonlinear problems, including equations of the Monge-Ampère type. Presently, computing reliable numerical freeform solutions to many such problems with high accuracy is possible but can be hard in terms of the complexity of the algorithms and computing time requirements. Because of the underlying strong nonlinearities, a reasonably accurate discretization of such design problems requires fine meshes for representing the light source and its radiation pattern, the target set, the required irradiance on the target, and the solution. Consequently, the algorithms and their implementations into codes may be quite complex as they have to solve a large number of equations relating the data on the source with the data on the target. Thus, many theoretical and computational issues important for applications still remain open. In particular, analysis and enhancements of computational algorithms for solving Monge-Ampère equations numerically remains an important direction of research. This direction of mathematical and optical research has become very active, especially during the last two decades (Oliker & Prussner, 1988; Oliker, 1989; Caffarelli et al., 1999; Glowinski et al., 2018; Nochetto & Zhang, 2018; de Castro et al., 2016).

The selection of problems considered in this chapter was aimed at providing illustrations of applications and design methods. Consequently, many applications which may benefit from the use of freeform surfaces and developed design methods are not discussed here. In order to focus our presentation and due to constraints on the size of this discussion we discuss only problems involving reflecting optical system. The refractive systems will be discussed in a subsequent publication. Chapter 8 is partially based on our earlier survey (Oliker, 2003). In order to maintain a reasonable level of rigor, we had to use some mathematical terminology. For the convenience of the reader, a summary of notations and terminology is provided at the beginning of this chapter. Beginning graduate textbooks (Lieb & Loss, 1997; Taylor, 2006) can be used as general references. Finally, we note the paper of Fang et al. (2013) which includes an extensive list of references to papers where the manufacturing of freeform optics is discussed.

REFERENCES

Benítez, P., Winston, R., and Miñano, J. C. (2006). *Nonimaging Optics*. Short Course # SC388, 76–78.

Berry, M. V. (2017). Laplacian magic windows. *J. Opt.* **19**, 1–5.

Bösel, C., and Gross, H. (2017). Single freeform surface design for prescribed input wavefront and target irradiance. *J. Opt. Soc. Am. A* **34**(9), 1490–1499.

Burkhard, D. G., and Shealy, D. L. (1987). A different approach to lighting and imaging: Formulas for flux density, exact lens and mirror equations and caustic surfaces in terms of the differential geometry of surfaces. In *SPIE Vol. 692 Materials and Optics for Solar Energy Conversion and Advanced Lighting Technology*, 248–272.

Caffarelli, L. A., and Oliker, V. (2008). Weak solutions of one inverse problem in geometric optics. Preprint, 1994. *J. Math. Sci.* **154**(1), 37–46.

Caffarelli, L. A., Kochengin, S., and Oliker, V. (1999). On the numerical solution of the problem of reflector design with given far-field scattering data. *Contemp. Math.* **226**, 13–32.

de Castro, P. M. M., Mérigot, Q., and Thibert, B. (2016). Far-field reflector problem and intersection of paraboloids. *Numer. Math.* **134**, 389–411.

Dickey, F. M. (Ed.) (2014). *Laser Beam Shaping: Theory and Techniques*, 2nd Ed. Taylor & Francis, Boca Raton, FL.

Dickey, F. M., Holswade, S., and Shealy, D. L. (2005). *Laser Beam Shaping Applications*. Taylor & Francis, Boca Raton, FL.

Fang, F. Z., Zhang, X. D., Weckenmann, A., Zhang, G. X., and Evans, C. (2013). Manufacturing and measurement of freeform optics. *CIRP Ann.-Manuf. Technol.* **62**, 823–846.

Galindo-Israel, V., Imbriale, W. A., and Mittra, R. (1987). On the theory of the synthesis of single and dual offset shaped reflector antennas. *IEEE Trans. Antennas Propag.* **AP-35**(8), 887–896.

Galindo-Israel, V., Imbriale, W. A., Mittra, R., and Shogen, K. (1991a). On the theory of the synthesis of offset dual-shaped reflectors – case examples. *IEEE Trans. Antennas Propag.* **39**(5), 620–626.

Galindo-Israel, V., Mittra, R., and Cha, A. G. (1979). Aperture amplitude and phase control on offset dual reflectors. *IEEE Trans. Antennas Propag.* **AP-27**, 154–164.

Galindo-Israel, V., Rengarajan, S., Imbriale, W. A., and Mittra, R. (1991b). Offset dual-shaped reflectors for dual chamber compact ranges. *IEEE Trans. Antennas Propag.* **39**(7), 1007–1013.

Glimm, T., and Oliker, V. (2003). Optical design of single reflector systems and the Monge-Kantorovich mass transfer problem. *J. Math. Sci.* **117**(3), 4096–4108.

Glimm, T., and Oliker, V. (2004). Optical design of two-reflector systems, the Monge-Kantorovich mass transfer problem and Fermat's principle. *Indiana Univ. Math. J.* **53**(5), 1255–1278.

Glowinski, R., Liu, H., Leung, S., and Qian, J. (2018). A finite element/operator- splitting method for the numerical solution of the two dimensional elliptic Monge-Ampère equation. *J. Sci. Comput.* **79**(1), 1–47.

Graf, T., and Oliker, V. I. (2012). An optimal mass transport approach to the near-field reflector problem in optical design. *Inverse Probl.* **28**(2), 1–15.

Kinber, B. E. (1962). On two reflector antennas. *Radio Eng. Electron. Phys.* **7**(6), 973–979.

Kinber, B. E. (1984). *Inverse Problems of the Reflector Antennas Theory - Geometric Optics Approximation.* preprint No. 38, Academy of Sc., USSR, in Russian.

Kochengin, S., and Oliker, V. (1997). Determination of reflector surfaces from near-field scattering data. *Inverse Probl.* **13**(2), 363–373.

Kochengin, S., and Oliker, V. (1998). Determination of reflector surfaces from near-field scattering data II. Numerical solution. *Numer. Math.* **79**(4), 553–568.

Kochengin, S., and Oliker, V. (2003). Computational algorithms for constructing reflectors. *Comput. Visualization Sci.* **6**, 15–21.

Kochengin, S., Oliker, V., and von Tempski, O. (1998). On the design of reflectors with pre-specified distribution of virtual sources and intensities. *Inverse Probl.* **14**, 661–678.

Lieb, E. H., and Loss, M. (1997). *Analysis.* Graduate Studies in Mathematics, v. 14, AMS.

Malyak, P. H. (1992). Two-mirror unobscured optical system for reshaping irradiance distribution of a laser beam. *Appl. Opt.* **31**(22), 4377–4383.

Newman, E., and Oliker, V. (1994). Differential-geometric methods in design of reflector antennas. In *Symposia Mathematica*, volume 35, pages 205–223.

Nochetto, R. H., and Zhang, W. (2018). Pointwise rates of convergence for the Oliker-Prussner method for the Monge-Ampère equation. *Numer. Math.* **141**(1), 253–288.

Oliker, V. (1989). On reconstructing a reflecting surface from the scattering data in the geometric optics approximation. *Inverse Probl.* **5**, 51–65.

Oliker, V. (2003). Mathematical aspects of design of beam shaping surfaces in geometrical optics. In *Trends in Nonlinear Analysis*, ed. by M. Kirkilionis, S. Krömker, R. Rannacher, and F. Tomi, pp. 193–224, Springer-Verlag, Berlin.

Oliker, V. (2005a). Geometric and variational methods in optical design of reflecting surfaces with prescribed irradiance properties. In *SPIE Proceedings, Nonimaging Optics and Efficient Illumination Systems II*, ed. by R. Winston and R. J. Koshel, volume 5942, pages 07-1–07-12, San Diego, CA.

Oliker, V. (2005b). Optical design of two-mirror beam-shaping systems. Convex and non-convex solutions for symmetric and non-symmetric data. In *SPIE Proceedings, Laser Beam Shaping VI - Optical Engineering and Instrumentation*, ed. by F. M. Dickey and D. L. Shealy, volume 5876, pages 203–214, San Diego, CA.

Oliker, V. (2006a). A rigorous method for synthesis of offset shaped reflector antennas. *Comput. Lett.* **2**(1–2), 29–49.

Oliker, V. (2006b). Freeform optical systems with prescribed irradiance properties in the near-field. In *SPIE Proceedings, International Optical Design Conference*, ed. by G. G. Gregory, J. M. Howard, and R. J. Koshel, volume 6342, pages 11-1–11-12, Vancouver, BC, Canada.

Oliker, V. (2007). Optical design of freeform two-mirror beam-shaping systems. *J. Opt. Soc. Am. A* **24**(12), 3741–3752.

Oliker, V. (2011). Designing freeform lenses for intensity and phase control of coherent light with help from geometry and mass transport. *Arch. Rational Mech. Anal.* **201**, 1013–1045.

Oliker, V. (2017). Controlling light with freeform multifocal lens designed with supporting quadric method (SQM). *Opt. Express*, **25**(4), A59–A72.

Oliker, V. I. (2014). Differential equations for design of a freeform single lens with prescribed irradiance properties. *Opt. Eng.* **53**(3), 031302-1–10.

Oliker, V., and Prussner, L. D. (1988). On the numerical solution of the equation and its discretizations, I. *Numer. Math.* **54**, 271–293.

Oliker, V., and Prussner, L. D. (1994). A new technique for synthesis of offset dual reflector systems. In *10-th Annual Review of Progress in Applied Computational Electromagnetics*, pages 45–52.

Oliker, V., Prussner, L. D., Shealy, D., and Mirov, S. (1994). Optical design of a two-mirror asymmetrical reshaping system and its application in superbroadband color center lasers. In *SPIE Proceedings*, volume 2263, pages 10–18, San Diego, CA.

Rhodes, P. W., and Shealy, D. L. (1980). Refractive optical systems for irradiance redistribution of collimated radiation: Their design and analysis. *Appl. Opt.* **19**(20), 3545–3553.

Ries, J., and Muschaweck, J. (2002). Tailored freeform optical surfaces. *J. Opt. Soc. Am. A* **19**(3), 590–595.

Schruben, J. S. (1974). Analysis of rotationally symmetric reflectors for illuminating systems. *J. Opt. Soc. Am.* **64**(1), 55–58.

Taylor, M. E. (2006). *Measure Theory and Integration*. Graduate Studies in Mathematics, v. 76, AMS.

Westcott, B. S. (1983). *Shaped Reflector Antenna Design*. Research Studies Press, Letchworth, England.

Westcott, B. S., and Norris, A. P. (1975). Reflector synthesis for generalized far fields. *J. Phys. A: Math. Gen.* **8**, 521–532.

Winston, R., Miñano, J. C., and Benítez, P., with contributions by Shatz, N., and Bortz, J. (2005). *Nonimaging Optics*. Elsevier Academic Press, Amsterdam.

8 Supporting Quadric Method (SQM)

8.1 PRECISE STATEMENT OF THE FFR PROBLEM

First we will complete the precise formulation of the far-field reflector problem which we began to discuss at the beginning of the introduction in Chapter 7. We will also continue to refer to Figure 7.1.

For the light ray of direction m from the source O incident on reflector R at point $r(m)$ with normal $u = u(m)$ the reflected direction y is determined by the law of reflection as

$$y = m - 2\langle m, u \rangle u, \tag{8.1}$$

where $\langle m, u \rangle$ is the dot product of m and u in $\mathbb{R}^3$. Thus, the surface R defines a map $\gamma: m \to y$ which maps the input aperture $\bar{D} \subset \mathbb{S}^2$ onto some set $\bar{T} \subset \mathbb{S}^2$; see Figure 7.1, where for convenience the input aperture $\bar{D}$ and the set of reflected directions $\bar{T}$ (the "output aperture" in the "far-field") are shown on the same sphere $\mathbb{S}^2$. According to the reflection law of geometric optics (Born & Wolf, 1999) in infinitesimal form, the irradiance produced by light reflected in direction $y = \gamma(m)$ is given by

$$\frac{I(m)}{|\det J(\gamma(m))|}$$

where $J(\gamma)$ denotes the matrix of the Jacobian of the map γ and det is the determinant (assuming that $\det J(\gamma) \neq 0$).

Suppose now that the sets D and T on $\mathbb{S}^2$ are given as well as nonnegative and integrable functions $I(m)$ on D and $L(y)$ on T. The FFR problem is to find a reflecting surface ("reflector") R such that for the map γ defined by R the following conditions are satisfied:

$$\gamma(\bar{D}) = \bar{T}, \tag{8.2}$$

$$L(\gamma(m)) \det J(\gamma(m)) = I(m), m \in D. \tag{8.3}$$

Because the functions L and I are assumed to be nonnegative, Condition (8.3) implies that it is assumed that $\det J(\gamma)(m)) \geq 0$. Conditions (8.2), (8.3) can be transformed into PDEs in terms of the radial (=sag) function $\rho(m)$, $m \in \bar{D}$, giving the distance from the origin O to the reflector R along the ray in direction m as shown in Figure 7.1 (cf. Schneider, 2014, p. 57). We refer the reader to our papers (Newman & Oliker, 1994; Oliker & Newman, 1993; Oliker & Waltman, 1987) for the detailed calculations leading to the formulas presented below.

The domain $\bar{D} \subset \mathbb{S}^2$ is described by a vector function $m(u^1, u^2)$ in some local coordinates u^1 and u^2 on the sphere $\mathbb{S}^2$. Usually, $u^1 = \theta$, $u^2 = \phi$ are the spherical coordinates but in some situations other coordinates may be more convenient. The formulas below are given in arbitrary coordinates u^1 and u^2 on $\mathbb{S}^2$. The input radiance $I(m) \equiv I(u^1, u^2)$ of the point source O is specified as a function of the source ray direction $m \in D$. The position vector of the reflector R to be determined by (8.2)–(8.3) is sought as a vector function $r(m) = \rho(m)m$, $m \in \bar{D}$. Thus, the goal is to determine the function $\rho(m)$, $m \in \bar{D}$.

We now transform the general Expressions (8.2) and (8.3) into expressions in terms of the function ρ. For that it is convenient to use some terminology from vector calculus and differential geometry. The unit length normal vector $u(m)$ on the surface R at the point $r(m) = \rho(m)m$ is given by

$$u = \frac{\rho m - \nabla \rho}{\sqrt{\rho^2 + |\nabla \rho|^2}}, \qquad (8.4)$$

where

$$\nabla \rho \equiv \mathrm{grad}\rho = \sum_{i,j=1}^{3} e^{ij} \frac{\partial \rho}{\partial u^i} \frac{\partial \rho}{\partial u^j}, \quad |\nabla \rho|^2 = \langle \nabla \rho, \nabla \rho \rangle,$$

with $[e^{ij}]$ being the inverse of the matrix $[e_{ij} \equiv e_{ij}(m)]$ of the first fundamental form ($\equiv$ metric) $ds^2 = \sum_{i,j=1}^{2} e_{ij} du^i du^j$ of the unit sphere $\mathbb{S}^2$,

$$e_{ij} = \left\langle \frac{\partial m}{\partial u^i}, \frac{\partial m}{\partial u^j} \right\rangle, \quad i, j = 1, 2.$$

Here and elsewhere, when no confusion may arise, in order to simplify the formulas we omit m when it is used as an argument of a function. The unit vector y in direction of the ray reflected off R for the incidence direction m can be expressed in terms of ρ and its derivatives upon substitution of Equation (8.4) into Equation (8.1). Then

$$y = \gamma(m) = m - 2 \frac{\rho(\rho m - \nabla \rho)}{|\nabla \rho|^2 + \rho^2}. \qquad (8.5)$$

The determinant of the Jacobian of the map γ is given by

$$\det J(\gamma) = \frac{\det\left[2\rho \nabla_{ij}\rho - \left(\rho^2 - |\nabla \rho|^2\right) e_{ij} - 4\rho_i \rho_j \right]}{\left(\rho^2 + |\nabla \rho|^2\right)^2 \det\left[e_{ij}\right]}, \qquad (8.6)$$

where

$$\nabla_{ij}\rho = \frac{\partial^2 \rho}{\partial u^i \partial u^j} - \sum_{k=1}^{2} \Gamma_{ij}^k \frac{\partial \rho}{\partial u^k}, \quad i, j, k = 1, 2,$$

and Γ_{ij}^k denote the Christoffel symbols of the second kind constructed from e_{ij} (see any elementary book on differential geometry). The expressions $\nabla_{ij}\rho$ are usually referred to as second covariant derivatives of the function ρ.

Now the FFR problem can be stated as follows. Find a function $\rho \in C^2(D) \cap C^1(\bar{D})$ (that is, ρ is one time continuously differentiable in $\bar{D}$ and twice continuously differentiable in D) such that the map γ defined by Equation (8.5) satisfies Equations (8.2) and (8.3), where $\det J(\gamma)$ is defined by (8.6).

The expression in Equation (8.6) involves second derivatives of the unknown function ρ under the determinant. Due to this special feature, Equation (8.3) and other equations of this type are referred to as PDEs of the Monge-Ampère type. The strong nonlinearities in Equations (8.5) and (8.6) (and consequently in Equations (8.2)–(8.3)) make a complete mathematically rigorous analysis of such equations very difficult. At the same time, most of the prescribed irradiance problems arising in optics and other fields require theoretical and numerical analysis of equations of the Monge-Ampère type.

Heuristic ad hoc (usually iterative) numerical approaches to specific problems of this kind have been used, often with only a posteriori graphic justification; see, for example, Kinber (1962); Kinber (1984); Galindo-Israel et al. (1979); Galindo-Israel et al. (1987); Galindo-Israel et al. (1991b); Westcott (1983); Westcott and Norris (1975); Ries and Muschaweck (2002); Brenner (2010); and Wu et al. (2013).

The FFR problem in various formulations has been studied in optics, mathematics, and numerical analysis (Keller, 1958; Kinber, 1962; Kinber, 1984; Galindo-Israel et al., 1979; Galindo-Israel et al., 1987; Galindo-Israel et al., 1991; Galindo-Israel et al., 1991; Westcott, 1983; Westcott & Norris, 1975; Caffarelli & Oliker, 2008; Caffarelli et al., 1999; Xu-Jia Wang, 1996; Guan & Wang, 1998; Caffarelli et al., 2008; Loeper, 2011; Fournier, 2010; Fournier et al., 2009; Oliker, 2002; Oliker, 2008; Glimm & Oliker, 2003; Mérigot & Oudset, 2016; de Castro et al., 2016; Wang, 2004; Romijn et al., 2019). Under a priori assumptions that the data and the solution are rotationally symmetric, the FFR problem was studied in Schruben (1974); Malyak (1992); Burkhard and Shealy (1987); and Oliker and Waltman (1987); see also Oliker (1987) and Oliker (1996). Geometric properties of reflectors have been studied in Newman and Oliker (1994); Hasanis and Koutroufiotis (1985); and Oliker, (2008a).

8.2   RELAXED FORMULATION, SQM, AND FREEFORM SURFACES

It turns out that if, in the formulation of the FFR problem, the requirement that $\rho \in C^2(D) \cap C^1(\bar{D})$ is relaxed and Equations (8.2)–(8.3) are replaced by more general but still physically meaningful relations, then a mathematically rigorous theory for solving an appropriately modified version of the FFR problem can be developed and a reliable solution in the class of **weak** solutions (explained in Section 8.6) can be obtained. This was done in the case of the two-reflector problem for the first time in Oliker and Prussner (1994) and an application of this result has been discussed already in Oliker et al. (1994). For single reflectors this was first done in Caffarelli and Oliker (2008). It is important to note that if there exists a $\rho \in C^2(D) \cap C^1(\bar{D})$ satisfying Equations (8.2)–(8.3) (in this case ρ is called a **strong** solution) then such ρ is also a weak solution. Thus, the class of weak solutions includes strong solutions of the FFR problem.

The concept of weak solutions was introduced in a geometric framework which allows a rigorous formulation of and solution to the FFR problem (Caffarelli and

Oliker, 2008; Caffarelli et al., 1999) and many other optical design problems (Oliker, 2006a; Kochengin and Oliker, 1998; Glimm and Oliker, 2003; Glimm and Oliker, 2004). In this framework any reflector solving the FFR problem is defined as an envelope of a family of paraboloids each of which is touching ("supporting") the reflector so that optical properties of paraboloids can be transferred to that reflector. This particular feature plays a crucial role in our considerations. In other optical problems the required optical surfaces are constructed as envelopes of families of "supporting" ellipsoids (Kochengin and Oliker, 1998; Kochengin and Oliker, 1997), hyperboloids (Oliker, 2017; Oliker, 2011; Oliker et al., 2015), or Cartesian ovals (Michaelis et al., 2011). This fact motivated our choice of the name for this framework as the **supporting quadric method** (SQM). Of course, Cartesian ovals are not quadrics but the method of design in this case remains the same. The reader will see that this framework is highly versatile as, in combination with the concept of weak solutions, it can be used to formulate, analyze, and solve a large variety of optical design problems concerned with the redirecting and redistribution of light.

Moreover, the solutions obtained with the SQM are **freeform surfaces**, that is, surfaces determined by the goals and constraints of the design without any a priori assumptions on their shape. In particular, freeform surfaces are not required to have any special symmetry, such as rotational or rectangular.

8.3 REFLECTORS AND RADIAL FUNCTIONS

We begin by redefining a reflector in purely geometric terms. For each $y \in \bar{T}$ we denote by $P(y)$ a paraboloid of revolution with focus at the origin O, axis of direction y—directed toward the opening of $P(y)$—and polar radius

$$\rho_y(m) = \frac{p(y)}{1 - \langle m, y \rangle}, \quad m \in \mathbb{S}^2 \setminus \{y\}, \tag{8.7}$$

where $p(y) > 0$ is the focal parameter of $P(y)$. For a given paraboloid $p(y) = \text{const} = \rho_y(m')$ when $m' \perp y$. If $p(y) = 0$ the paraboloid $P(y)$ is considered to be a limit of co-axial paraboloids with focal parameters tending to zero. Obviously, this limit is a half line originating at O.

Definition 8.3.1 *Suppose $\bar{T} \subset \mathbb{S}^2$ and contains more than one point. Let $\{P(y)\}_{y \in \bar{T}}$ be an arbitrary family of paraboloids with $p(y) > 0$ for all y. For each $y \in \bar{T}$ let $B(y)$ be the infinite convex solid bounded by $P(y)$. The surface R defined as*

$$R = \partial B, \quad \text{where} \quad B = \bigcap_{y \in \bar{T}} B(y), \tag{8.8}$$

is called a reflector defined by the family $\{P(y)\}_{y \in \bar{T}}$ with the light source at the (common) focus O. See Figure 8.1 for an illustration in a plane. The set of all reflectors with the source O is denoted by $\mathcal{R}$. An important special case is when $\bar{T} = \{y_1, y_2, \ldots, y_k\}, k \geq 2$. In this case we say that $\mathcal{R}$ is defined by paraboloids $P(y_i), i = 1, 2, \ldots, k$.

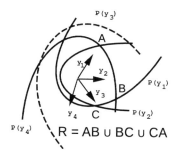

FIGURE 8.1 $\bar{T} = \{y_1, y_2, y_3, y_4\}$. Reflector R consists of the arcs AB, BC, and CA which are pieces of parabolas $P(y_i)$, $i = 1, 2, 4$; parabola $P(y_3)$ is not touching R. The set $\bar{D}$ and $\mathbb{S}^2$ are not shown.

Let $R \in \mathcal{R}$ and $\bar{T}, \rho$, be as they are in the above definition. Since $\bar{T}$ has more than one point, R is bounded. The solid $B(y)$ is convex for each $y \in \bar{T}$ and because the intersection of convex sets is convex, the set B is convex. Since $p(y) > 0$ for all $y \in \bar{T}$, the point O is in the interior of B. Thus, R is a bounded, closed (that is, $\partial R = \varnothing$) convex surface, star-shaped relative to O. It projects 1-1 onto $\mathbb{S}^2$ by rays from O.

Equivalently, if $R \in \mathcal{R}$ is defined by paraboloids $\{P(y)\}_{y \in \bar{T}}$, then

$$\rho(m) = \inf_{y \in \bar{T}} \frac{p(y)}{1 - \langle m, y \rangle} \equiv \inf_{y \in \bar{T}} \rho_y(m), \quad m \in \mathbb{S}^2, \tag{8.9}$$

and the $\inf_{y \in \bar{T}}$ in (8.9) is attained at least for one $y \in \bar{T}$. Thus, $\rho(m)$ is a specialization to reflectors in $\mathcal{R}$ of the notion of a **radial** function used in Section 8.1. Because any $R \in \mathcal{R}$ is closed and convex, its radial function is well defined on the entire $\mathbb{S}^2$ and R is the graph of ρ over $\mathbb{S}^2$.

It may happen that some of the paraboloids in the family $\{P(y)\}_{y \in \bar{T}}$ have no common points with R (as in the case of $P(y_3)$ in Figure 8.1). We want to exclude from consideration such paraboloids. More generally, in order to identify all paraboloids $P(y)$, $y \in \mathbb{S}^2$, "touching R from outside," we introduce the following

Definition 8.3.2 *Let $R \in \mathcal{R}$ and let ρ be the radial function of R. Let $y \in \mathbb{S}^2$ and let $P(y)$ be a paraboloid of revolution with focus O, axis of direction y, and radial function ρ_y. If*

$$\rho(m) \le \rho_y(m) \ \forall m \in \mathbb{S}^2, \tag{8.10}$$

$$\text{and } \rho(m) = \rho_y(m) \text{ for some } m \in \mathbb{S}^2, \tag{8.11}$$

then we say that $P(y)$ is supporting to R at the point $r(m) = \rho(m) m \in R$.

Note that $P(y)$ is not required to be in the family $\{P(y)\}_{y \in \bar{T}}$ defining R.

In Figure 8.1 the parabolas $P(y_1)$, $P(y_2)$ and $P(y_4)$ are supporting R at each point of the arcs AB, BC and CA, respectively, while $P(y_3)$ is not supporting to R. It follows from Equation (8.7) that by decreasing the focal parameter $p(y_3)$ of $P(y_3)$ to a suitable value $p'(y_3)$ a new parabola $P'(y_3)$ supporting to R is obtained. This property is applicable in much more general circumstances and it will play an important role in the construction of a reflector solving the FFR problem.

8.4 FOCAL FUNCTION

The above example shows that a reflector $R \in \mathcal{R}$ may have supporting paraboloids which are not among paraboloids in the family defining R. In fact, for any reflector $R \in \mathcal{R}$ it is possible to construct a paraboloid of revolution with any given axis y and focus O supporting to R. This is a consequence of the boundedness and closeness of R. The proof of this fact is nearly the same as in the example above. That is, if $P(y)$ is not already supporting to R, rescale $P(y)$ homothetically into $P'(y)$ so that $\rho(m) \le \rho'_y(m)$ $\forall m \in \mathbb{S}^2$, where ρ'_y is the radial function of $P'(y)$. Then decrease the focal parameter $p'(y)$ until $P'(y)$ is supporting to R. Because R is a closed bounded surface, the number p' can always be uniquely determined.

Based on this property, applicable to any $y \in \mathbb{S}^2$ and any reflector $R \in \mathcal{R}$, it is always possible to define for any $R \in \mathcal{R}$ a function $p(y)$, $y \in \mathbb{S}^2$, such that each paraboloid $P(y)$ with focus O and focal parameter $p(y)$ is supporting to R. This function is called the *focal function of reflector R*. Note that $p(y) > 0$ $\forall y \in \mathbb{S}^2$, since the origin O is strictly inside the convex body bounded by R.

Using the focal function, Conditions (8.10) and (8.11) can be restated as follows: for each $y \in \overline{T}$ for which the $\inf_{\overline{T}}$ on the right of Equation (8.9) is achieved, that is, $P(y)$ is supporting, we have

$$p(y) \ge \rho(m)(1 - \langle m, y \rangle) \quad \text{for all } m \in \mathbb{S}^2, \text{ and}$$

$$p(y) = \rho(m)(1 - \langle m, y \rangle) \quad \text{for all } m \text{ satisfying} \quad (8.11).$$

Therefore, for each $y \in \overline{T}$

$$p(y) = \sup_{m \in \mathbb{S}^2} \left[\rho(m)(1 - \langle m, y \rangle) \right]. \tag{8.12}$$

Clearly, the equality in Equation (8.12) holds for any paraboloid supporting to R, including those with axes $y \in \mathbb{S}^2 \setminus \overline{T}$. Thus, if $R \in \mathcal{R}$ and if $\rho(m)$, $m \in \mathbb{S}^2$ and $p(y)$, $y \in \mathbb{S}^2$ are the radial and focal functions of R, respectively, then

$$\rho(m) = \inf_{y \in \mathbb{S}^2} \frac{p(y)}{1 - \langle m, y \rangle} \quad \forall m \in \mathbb{S}^2, \tag{8.13}$$

$$p(y) = \sup_{m \in \mathbb{S}^2} \left[\rho(m)(1 - \langle m, y \rangle) \right] \quad \forall y \in \mathbb{S}^2. \tag{8.14}$$

It can be shown that for any $R \in \mathcal{R}$ the functions ρ, $p \in C(\mathbb{S}^2)$. The continuity of ρ is a consequence of convexity of R. Also, since R is bounded, the function ρ is bounded. The continuity of p follows from Equation (8.14) and the boundedness of ρ. Thus, with each reflector $R \in \mathcal{R}$ we have a uniquely defined pair of continuous on $\mathbb{S}^2$ functions p, $\rho > 0$ satisfying Equations (8.13)–(8.14). Conversely, any pair of continuous and positive functions on $\mathbb{S}^2$ satisfying Equations (8.13)–(8.14) defines uniquely a reflector in $\mathcal{R}$. Indeed, the family of paraboloids defined by $p(y)$, $y \in \mathbb{S}^2$, can be used as the family defining some $R \in \mathcal{R}$ in the construction (8.8). At the same time, by Equation (8.14) each paraboloid in this family is supporting to R. Clearly, such R is defined by ρ and p uniquely.

Note 8.4.1 It follows from the preceding considerations that, while we are dealing with the radial and focal functions defining a reflector $R \in \mathcal{R}$, we may assume that the sets $\bar{D}$ and $\bar{T}$ coincide with the sphere $\mathbb{S}^2$. However, as we will see in the following sections one has to deal carefully with such assumptions if considerations involve the radiance and irradiance over the corresponding input and output apertures.

8.5   GENERALIZED REFLECTOR MAP

Next, we want to extend the notion of reflector map to reflectors in $\mathcal{R}$. In general, a reflector $R \in \mathcal{R}$ may have points at which there is no tangent plane and the normal is not defined. Therefore, the reflected direction can not be defined using the reflection law (8.1). For example, in Figure 8.1 such points are A, B, and C. Note that at such a point there is more than one supporting paraboloid. Nevertheless, we can use Equation (8.9) to define an analogue of the reflector map, called the *generalized* reflector map.

Let R be an arbitrary reflector in $\mathcal{R}$. For a point $m \in \mathbb{S}^2$ the vector $r(m) = \rho(m)m$ gives the corresponding point on R. It follows from Equation (8.9) that the inf is achieved for $P(y)$ supporting to R at $r(m)$. *The image of m under the generalized reflector map—we denote it again by $\gamma(m)$—is defined as*

$$\gamma(m) = \left\{ y \in \bar{T} \mid p(y) = \rho(m)(1 - \langle m, y \rangle) \right\}, \quad m \in \mathbb{S}^2. \tag{8.15}$$

According to this formula, if at the point $r(m)$ there is a unique supporting paraboloid then the unit vector in the direction of its axis is $\gamma(m)$. However, if at the point $r(m)$ there is more than one supporting paraboloid then $\gamma(m)$ is the set of *all* unit vectors in $\mathbb{S}^2$ in directions of axes of the corresponding supporting paraboloids. Thus, in general, γ is multivalued. It is also clear that the map g maps $\mathbb{S}^2$ onto $\bar{T}$, as for each $y \in \bar{T}$ there is a paraboloid with axis y supporting to R. If we include in our consideration all possible supporting paraboloids to R then clearly $\gamma(\mathbb{S}^2) = \mathbb{S}^2$.

The points on a reflector R at which there is more than one supporting paraboloid are called *singular*. The nonsingular points are called *smooth*. Thus, a ray emitted from O in direction $m \in \mathbb{S}^2$ is reflected off R at $r(m)$ and produces a *a set of directions* $\gamma(m)$. If $r(m)$ is a smooth point on R then $g(m)$ consists just of one, uniquely defined, point. On the other hand, if $r(m)$ is a singular point then this ray splits into a set of rays with directions $\gamma(m)$. This is consistent with the physical interpretation of diffraction. In this case, γ is multivalued and, clearly, $\bar{T} \subset \gamma(\mathbb{S}^2)$.

It turns out that the set of singular points on R is "small" in the sense explained later in Notes 5.1–5.2. Observe also that expanding the family of supporting paraboloids to R by including supporting paraboloids with axes of direction $y \in \mathbb{S}^2 \setminus \bar{T}$ leaves the the radial function of R unchanged. It is also clear that a paraboloid defined by $p(y)$ with $y \in \mathbb{S}^2 \setminus \overline{T}$ can be supporting to R only at a singular point. Also, a supporting paraboloid $P(y)$ is tangent to R at a smooth point only if $y \in T$. It can be shown that any $R \in \mathcal{R}$ is completely determined by supporting paraboloids tangent to R at smooth points; cf. Schneider (Schneider, 2014), Theorem 2.2.6.

Note that each tangent plane to a supporting paraboloid at the point of contact with R leaves the paraboloid and the reflector in one of the half-spaces defined by that plane; that is, this plane is supporting to R in the sense of convexity theory (Schneider, 2014). Thus, any singular point on R will also be singular in the sense of convexity theory, that is, at such a point the surface R has more than one supporting plane.

Note 8.5.1 Since reflectors in $\mathcal{R}$ are convex surfaces, it is useful to recall some basic facts from convexity theory. These facts will often be used in this exposition. An extensive source for theory of convex surfaces is Schneider (2014). We also use here some basic notions of measure theory; see, for example, Lieb and Loss (1997) and Taylor (2006). Let $R \in \mathcal{R}$. On any convex surface it is possible to define Borel sets obtained from closed subsets of R by taking countable unions and intersections. Each Borel set on R has a well-defined (Lebesgue) measure (area) and we often refer to Borel sets on R as measurable. For a set $A \subset R$ we say that a property holds almost everywhere on A when it holds on $A \backslash A'$, where the measure of A' is zero. It turns out that any reflector $R \in \mathcal{R}$ is smooth almost everywhere. This follows from the fact that any tangent plane to a supporting paraboloid at the point of contact with R leaves the paraboloid and the reflector on one side defined by that plane. In convexity theory such planes are also called supporting. Thus, any singular point on R will also be singular in the sense of convexity theory because at such a point R has more than one supporting plane. By a known result in convexity theory the set of singular points on a convex surface has measure zero (Schneider, 2014, p. 89). Therefore, the set of singular points on a reflector in $\mathcal{R}$ also has measure zero.

Note 8.5.2 It is known from convexity theory that the set of points where a convex surface is not two times differentiable has measure zero (Schneider, 2014, p. 31). Thus, any $R \in \mathcal{R}$ is almost everywhere two times differentiable. From this we conclude, in particular, that the radial function defining a reflector is differentiable two times almost everywhere on $\mathbb{S}^2$. Furthermore, it was shown in Oliker (2002), on p. 160, that the focal function p of a reflector $R \in \mathcal{R}$ is also the support function of a closed, bounded convex surface in $\mathbb{R}^3$. By a known result in convexity theory (Schneider, 2014, p. 31), the focal function p is twice differentiable almost everywhere on R. The above observations imply also that for any $R \in \mathcal{R}$ the map $\gamma : \mathbb{S}^2 \to \mathbb{S}^2$ is single-valued almost everywhere on $\mathbb{S}^2$.

8.6   WEAK SOLUTIONS

In this section we work with input and output apertures $\overline{D}, \overline{T} \subset \mathbb{S}^2$, respectively. Each of the apertures may coincide with the entire $\mathbb{S}^2$. The radiation $I(m)$, $m \in \overline{D}$, of the source O is assumed to be nonnegative and integrable. Setting $I(m)=0$ for m in the complement of $\bar{D}$ in $\mathbb{S}^2$, we make I defined on the entire $\mathbb{S}^2$. In addition, $I \in L(\mathbb{S}^2)$. For a measurable set $\Omega \subset \mathbb{S}^2$ the quantity

$$\int_\Omega I(m)d\sigma(m),$$

where $d\sigma(m)$ is the surface area element of $\mathbb{S}^2$, is the total energy of the source emitted through the aperture Ω.

In order to define the amount of energy transferred by reflector R from the source O to a subset $\omega \subset \bar{T}$, we first need to introduce the *inverse* map $\gamma^{-1} : \bar{T} \to \mathbb{S}^2$ which associates with each $y \in \bar{T}$ the set of directions m of incident rays reflected by R in direction y. Analytically,

$$\gamma^{-1}(y) = \left\{ m \in \mathbb{S}^2 \mid p(y) = \rho(m)(1 - \langle m, y \rangle) \right\}, \quad y \in \bar{T}.$$

It was shown in Caffarelli and Oliker (2008) and, in more detail, in Oliker (2002) that for any measurable set $\omega \subset \bar{T}$ the set $\gamma^{-1}(\omega)$ is measurable on $\mathbb{S}^2$, and, therefore, the function

$$G(R,\omega) = \int_{\gamma^{-1}(\omega)} I(m)d\sigma(m), \tag{8.16}$$

is a completely additive measure defined on all Borel subsets of $\bar{T}$. Physically, $G(R, \omega)$ is the quantitative expression of the amount of energy transferred by reflector R to the far-field region defined by reflected rays with directions $y \in \omega$. Note that $G(R, \omega)$ may not be absolutely continuous; that is, $G(R, \omega)$ may be positive even if ω consists of only one point.

Let function L define the required irradiance on $\bar{T}$ be as before, that is, $L \in L(\bar{T})$ and nonnegative. Extend L to the entire $\mathbb{S}^2$ by setting $L(y)=0$ when $y \in \mathbb{S}^2 \setminus \bar{T}$. Thus, $L \in L(\mathbb{S}^2)$.

Definition 8.6.1 *A closed convex reflector* $R \in \mathcal{R}$ *is called a* **weak solution** *to the FFR problem if*

$$G(R,\omega) = \int_\omega L(y)d\sigma(y) \quad \text{for any measurable set} \quad \omega \subset \mathbb{S}^2. \tag{8.17}$$

Equation (8.17) for the unknown reflector R together with Equation (8.16) is in terms of integrated intensities and it is called the **weak** form of the corresponding pointwise equation

$$I(\gamma^{-1}(y))\left|J(\gamma^{-1}(y))\right| = L(y) \quad \text{on} \quad T, \tag{8.18}$$

which in turn is equivalent to Equation (8.3) if the latter is restated on T and treated as the target set under reflector map γ. Evidently, for an arbitrary $R \in \mathcal{R}$ such a transition is not possible and Equation (8.17) is a physically motivated replacement of Equation (8.18).

Example 1. Here we provide a physical illustration of Equation (8.17) and its weak solution. Suppose $\overline{T}$ consists of only two points $y_1, y_2, y_1 \neq y_2$ and

$$L(y) = A\delta(y - y_1) + 0\delta(y - y_2), \quad A = \int_D I(m)d\sigma(m),$$

where $\delta(y - y_i)$, $i = 1, 2$ is the Dirac delta function on $\mathbb{S}^2$ concentrated at y_i. The reflector R is constructed using two paraboloids: $P(y_1)$ with focal parameter p_1 and $P(y_2))$ with focal parameter $p_2 \gg p_1$. Assume that D is such that $P(y_1)$ intercepts all light rays emitted by O through $\overline{D}$. Then $g(m) = y_1$ for all $m \in \overline{D}$, $\det J(\gamma(m)) = 0$ everywhere in $\overline{D}$, and $\gamma^{-1}(y_1) = \overline{D}$. At the same time,

$$G(R, y_1) = \int_{\gamma^{-1}(y_1)} I(m)d\sigma(m) = A, \quad G(R, y_2) = 0,$$

which is to say, all reflected light is concentrated at y_1. Thus, Equation (8.17) is a physically correct description of this situation while the point-wise Equation (8.18) is not applicable.

Example 2. Suppose $D = \mathbb{S}^2$, $\overline{T} = \{y_1, y_2\}$, $y_1 \neq y_2$, $I(m) \equiv 1$, and

$$L(y) = F_1\delta(y - y_1) + F_2\delta(y - y_2), F_1, F_2 > 0, \text{ and } F_1 + F_2 = \int_{\mathbb{S}^2} I(m)d\sigma(m).$$

Again, our reflector R is constructed with two paraboloids of revolution with axes y_1 and y_2. Let p_1 and p_2 be two positive numbers which we use as focal parameters of $P(y_1)$ and $P(y_2)$, respectively. It is clear that if p_1 and take p_2 sufficiently large then for the corresponding reflector the area of the set $g^{-1}(y_2)$ will be small and consequently the integral

$$F(y_2) := \int_{g^{-1}(y_2)} I(m)d(m)$$

will have a small value; that is, the total irradiance from the source O reflected by R in direction y_2 will be small. The analogous integral $F(y_1)$ will be nearly equal to $F_1 + F_2$. By decreasing p_2 continuously we will obtain reflectors producing larger $F(y_2)$ and smaller $F(y_1)$. The required values of F_1 and F_2 will be achieved at some value of p_2. The corresponding pair of paraboloids combined as in Definition 3.1 is the weak solution to FFR in this case.

The following important property connects the energy G transferred by a reflector, the generalized reflector map γ, and the input radiance I.

Theorem 8.6.2 *For any reflector $R \in \mathcal{R}$ and any continuous function h on $\mathbb{S}^2$ the following equality holds:*

$$\int_{\mathbb{S}^2} h(y)G(R, d\sigma(y)) = \int_{\mathbb{S}^2} h(\gamma(m))I(m)d\sigma(m), \tag{8.19}$$

where γ is the generalized reflector map of R. The integral on the left is taken with respect to the measure $G(R, \omega)$ defined in Equation (8.16).

Because g is multi-valued only on a subset of $\mathbb{S}^2$ which has zero area, the integral on the right of Equation (8.19) is well defined. The proof of this property is analogous to the proof of Lemma 4.9 in Glimm and Oliker (2004).

8.7   WEAK SOLUTIONS TO THE FFR PROBLEM AND ITS SEMI-DISCRETE VERSION

In this section it is shown that existence of weak solutions to the FFR problem can be established rigorously under very general conditions (Caffarelli & Oliker, 2008). Put

$$F(\omega) = \int_\omega L(y) d\sigma(y) \quad \text{for any measurable set} \quad \omega \subset \mathbb{S}^2. \tag{8.20}$$

Theorem 8.7.1 *Let I and L be nonnegative integrable functions on $\mathbb{S}^2$ and*

$$\int_{\mathbb{S}^2} I(m) d\sigma(m) = \int_{\mathbb{S}^2} L(y) d\sigma(y) \neq 0. \tag{8.21}$$

Then the FFR problem in weak formulation admits a solution $R \in \mathcal{R}$ such that

$$G(R, \omega) = F(\omega) \quad \text{for any measurable set} \quad \omega \subset \mathbb{S}^2. \tag{8.22}$$

If $\inf_{\mathbb{S}^2} I > 0$ then this solution is unique up to homothety relative to O.

Remark 8.7.2 The measure F in Equation (8.20) is defined by a nonnegative integrable function L. This and Condition (8.21) imply that F is a positive Borel measure. In fact, Theorem 8.7.1 is valid for Borel measures not necessarily generated by nonnegative integrable functions. It suffices to assume that F is a positive Borel measure not concentrated at one point. On the other hand we note that if F is concentrated at one point, say $\bar{y}$, then any paraboloid with axis of direction $\bar{y}$ is obviously a solution to the FFR problem.

The requirement that F is not concentrated at one point is important as it allows us to conclude that the set of reflectors $\mathcal{R}$ in which the solution is found can be defined so that it is not empty and the radial and focal functions of all reflectors in $\mathcal{R}$ satisfy the inequalities

$$0 < c_1 = (m), \, p(y) = c_2 < \infty, \tag{8.23}$$

where the constants c_1, c_2 depend only on I and F; see Caffarelli and Oliker (2008).

The proof of Theorem 8.7.1 is sketched below and follows the proof in Caffarelli and Oliker (2008). It is important to sketch it here because it is based on an approximation scheme that can also be used for solving the problem numerically (this aspect is discussed later). In addition, the proof clearly illustrates the ideas behind the SQM. Theorem 8.7.1 is proved in two main steps. On the first step, the measure F is approximated by a sum of Dirac masses

$$F^k = \sum_{i=1}^{k} F_i^k \delta(y - y_i)$$

concentrated at a finite number of points $y_1^k, \ldots, y_k^k$ on $\mathbb{S}^2$. For each F^k the reflector problem is solved by a reflector determined by k paraboloids of revolution with a common focus at O. The corresponding FFR problem is called **semi-discrete**. On the second step, we let $k \to \infty$ and use weak convergence of measures F^k to G to obtain a weak solution to Equation (8.22). For the convenience of the reader we recall now the notion of weak convergence used in this work.

Let G and F_k, $k = 1, 2, \ldots$ be completely additive and nonnegative measures defined on Borel subsets of $\mathbb{S}^2$. If

$$\lim_{k \to \infty} \int_{\mathbb{S}^2} f(y) F^k(d\sigma) = \int_{\mathbb{S}^2} f(y) G(d\sigma)$$

for any continuous function f on $\mathbb{S}^2$ then it is said that the sequence $\{F^k\}$ converges weakly to G (Lieb & Loss, 1997; Taylor, 2006). The integrals on the left and on the right are taken with respect to measures F^k and G respectively.

We outline now the proof of Theorem 8.7.1.

8.7.1 DERIVATION OF THE SEMI-DISCRETE FFR PROBLEM

At this stage, for each $k \in \mathbb{N}$, $k \geq 2$ we subdivide $\mathbb{S}^2$ into k measurable subsets $\omega_1^k, \ldots, \omega_k^k$, such that for each $i = 1, \ldots, k$ the diameters $\omega_i^k < 3\pi / k$ and

$$\bigcup_{i=1}^{k} \bar{\omega}_i^k = \mathbb{S}^2, \quad \omega_i^k \cap \omega_j^k = \varnothing \quad \text{if } i \neq j.$$

Pick points $y_i^k \in \omega_i^k$, $i = 1, 2, \ldots, k$, and put

$$F_i^k = \int_{\omega_i^k} L(y) d\sigma(y), \quad i = 1, \ldots, k,$$

and assume that y_i^k are chosen so that $F_i^k > 0$ for at least two points. For a measurable set $\omega \subset \mathbb{S}^2$ we define

$$F^k(\omega) = \sum_{y_i^k \in \omega} F_i^k$$

Since the "balance" Equation (8.21) holds, the equation

$$\int_{\mathbb{S}^2} I(m) d\sigma(m) = F^k(\mathbb{S}^2) = \sum_{i=1}^{k} F_i^k \tag{8.24}$$

is also satisfied.

Fix some k as above. We want to find a reflector $R^k \in \mathcal{R}$ defined by paraboloids $P\left(y_i^k\right)$, $i = 1, \ldots, k$, with focal parameters $p_i^k > 0$, $i = 1, \ldots, k$, such that

$$G\left(R^k, y_i^k\right) = F_i^k, \quad i = 1, \ldots, k. \tag{8.25}$$

The problem of constructing the reflector $R^k \in \mathcal{R}$, satisfying Equation (8.25) with given $y_i^k, F_i^k, i = 1,\ldots,k$, function I as above, and $F_i^k, i = 1,\ldots,k$, satisfying Equation (8.24), is the explicit form of the **semi-discrete** FFR problem. We claim that the semi-discrete problem can be solved and the proof of existence of the required reflector is based on the following arguments.

8.7.2 SOLUTION TO EQUATION (8.25)

First, we note that any k paraboloids $P(y_i^k)$, $i = 1,\ldots,k$, of revolution with axes of direction y_i^k, foci at O and focal parameters $p_i^k > 0$ for each i form a family of paraboloids defining a reflector $R^k \in \mathcal{R}$ (as described in Section 8.2). For any such reflector we have

$$\sum_{i=1}^{k} G\left(R^k, y_i^k\right) = E, \quad \text{where for brevity we put } E = F^k(\mathbb{S}^2).$$

Second, the functions $G\left(R^k, y_i^k\right)$ are invariant relative to homotheties of R^k with respect to the origin O. Therefore, we can pick any such reflector and rescale it so that the focal parameter $p_1^k = 1$ for the paraboloid $P\left(y_1^k\right)$. Fix one such reflector and denote it by R_1^k. Relabeling the points y_i^k if necessary, we can index the points $y_1^k,\ldots,y_k^k$ so that $F_1^k > 0$. Next, observe that for $i > 1$ the function $G\left(R_1^k, y_i^k\right) \to 0$ when $p_i^k \to +\infty$. Therefore, the focal parameters $p_2^k > 0,\ldots,p_k^k > 0$ of paraboloids forming R_1^k can be chosen so large, and a number $f^k > 0$ so small, that

$$0 \le G\left(R_1^k, y_i^k\right) \le f^k \text{ for } i > 1 \text{ while}$$

$$G\left(R_1^k, y_1^k\right) = E - \sum_{i=2}^{k} G\left(R_1^k, y_i^k\right) \ge E - (k-1) f^k > 0.$$

At the same time, at least for one $i > 1$, the inequality $G\left(R_i^k, y_i^k\right) > 0$ holds; see Remark 7.2.

Consider now the paraboloid $P(y_2)$. If $G\left(R_1^k, y_2^k\right) < f^k$ we decrease p_2 while keeping the p_i for $i \neq 2$ fixed. As $p_2^k \to 0$ the function $G\left(R_1^k, y_2\right)$ is continuously increasing to a value as close to E as one wishes. Therefore, there exists a value $\overline{p}_2^k$ such that $G\left(R_2^k, y_2\right) = F_2^k$, where R_2^k denotes the reflector with paraboloid $\overline{P}\left(y_2^k\right)$ with focal parameter $\overline{p}_2^k$ replacing the original $P\left(y_2^k\right)$. During the process of decreasing p_2^k the $G\left(R_1^k, y_j\right)$ are nonincreasing for all $i \neq 2$. The same step is repeated for R_2^k and p_3^k and then for all remaining paraboloids in the family, except for $P\left(y_1^k\right)$. This process produces monotone sequences of $\overline{p}_i^k$, $i = 2,\ldots,k$ bounded away from zero and infinity. It is shown in Caffarelli and Oliker (2008) that the corresponding sequence of reflectors converges to a reflector in $\mathcal{R}$ for which equalities are achieved in Equation (8.25) for all i. This completes the proof of existence of solutions in the semi-discrete case.

To obtain the solution to the FFR problem in the general case, a sequence of subdivisions of $\mathbb{S}^2$ into subsets with diameters of $\omega_i^k \to 0$ as $k \to \infty$ for all $i = 1,\ldots,k$ is considered. These subdivisions are constructed as in the beginning of the proof. Consequently, we obtain an infinite sequence of reflectors $R^k, k = 2,\ldots$, with radial functions $\{\rho_1^k,\ldots,\rho_k^k\}$, $k \geq 2$, and focal parameters $\{p_1^k,\ldots,p_k^k\}$, $k \geq 2$. The estimate (8.23) allows us to conclude that there exists a subsequence of reflectors with radial functions denoted again by ρ^k, $k \geq 2$, converging in $C^0(\mathbb{S}^2)$ to a reflector $R \in \mathcal{R}$ with radial function $\rho(m) = \lim_{k\to\infty}\rho^k(m)$ for all $m \in \mathbb{S}^2$.

Next, it is shown that the measures $G(R^k,\cdot)$, $k \geq 2$ converge weakly to $G(R,\cdot)$, that is,

$$\lim_{k\to\infty}\int_{\mathbb{S}^2}h(y)G\left(R^k,d\sigma(y)\right) = \int_{\mathbb{S}^2}h(y)G\left(R,d\sigma(y)\right) = \int_{\mathbb{S}^2}h(\gamma(m))I(m)d\sigma(m) \quad (8.26)$$

for any function $h \in C(\mathbb{S}^2)$. On the other hand, the desired irradiance is

$$F(\omega) = \int_{\omega}L(y)d\sigma(y) \text{ for any } \omega \subset \mathbb{S}^2$$

and $F^1,\ldots,F^k$ are produced by discretization of F. Therefore,

$$\lim_{k\to\infty}\int_{\mathbb{S}^2}h(y)F^k\left(d\sigma(y)\right) = \int_{\mathbb{S}^2}h(y)F\left(d\sigma(y)\right) = \int_{\mathbb{S}^2}h(y)L(y)d\sigma(y). \quad (8.27)$$

Combining Equation (8.25) with Equations (8.26) and (8.27), we obtain

$$\int_{\mathbb{S}^2}h(y)L(y)d\sigma(y) = \int_{\mathbb{S}^2}h(y)G(R,d\sigma(y)) = \int_{\gamma^{-1}(\mathbb{S}^2)}h(\gamma(m))I(m)d\sigma(m)$$

for any contunuous h on $\mathbb{S}^2$. This implies Equation (8.17) and, therefore, R is the required weak solution. Again, further details can be found in Caffarelli and Oliker (2008).

In practical applications the FFR problem needs to be solved when the input and output apertures $\bar{D}$ and $\bar{T}$ do not coincide with $\mathbb{S}^2$. The construction of the weak solutions described above can be easily adapted to this formulation. Namely, for $\bar{D},\bar{T} \subset \mathbb{S}^2$ we extend the functions I and L to the entire $\mathbb{S}^2$ as before, by setting them equal to zero on $\mathbb{S}^2 \setminus \bar{D}$ and $\mathbb{S}^2 \setminus \bar{T}$, respectively. Since the balance Equation (8.21) is satisfied with $\bar{D}$ and $\bar{T}$, it will also be satisfied with the functions I and L extended to the entire $\mathbb{S}^2$. Then Theorem 8.7.1 can be applied. Next, we delete from the resulting closed reflector all of the supporting paraboloids with axes of directions in $\mathbb{S}^2 \setminus \bar{T}$. The part of reflector over $\mathbb{S}^2 \setminus \bar{D}$ can also be deleted, as the radiance of the source over that part is $\equiv 0$. This determines a reflector R' with a boundary whose radial projection on $\mathbb{S}^2$ is the set $\bar{D}$ and $\gamma(\bar{D}) = \bar{T}$. Evidently,

$$G(R', \omega) = \int_{\gamma^{-1}(\omega)} I(m) d\sigma(m) = \int_{\gamma^{-1}(\omega) \cap \bar{D}} I(m) d\sigma(m)$$

$$= \int_\omega L(y) d\sigma(y) \text{ for any measurable set } \omega \subset T,$$

that is, R' is the required reflector. The uniqueness of the solution in Theorem 8.7.1 was proved in Caffarelli and Oliker (2008) for reflectors defined by a finite number of paraboloids; the general case was treated in Guan and Wang (1998). For related results on intersection of paraboloids see de Castro et al. (2016).

8.8 NUMERICS AND SQM

The first provably convergent computational algorithm for solving the FFR problem was based on the SQM approach described in Section 8.7 for proving existence of weak solutions to Equation (8.17); see Caffarelli et al. (1999).

A weak solution is obtained as a limit of a sequence of convex reflectors $R^k \in \mathcal{R}$, $k = 2, 3, \ldots$, solving the corresponding semi-discrete Problems (8.25). Thus, for numerics, the first task is to determine numerical solutions of such semi-discrete problems.

For each k, a reflector R^k solving a semi-discrete FFR Problem (8.25) is defined by k supporting paraboloids with given axes $y_1^k, \ldots, y_k^k$ each of which produces irradiances $F_1^k, \ldots, F_k^k$ in corresponding directions. According to the procedure described in Section 8.7, the k supporting paraboloids with axes $y_1^k, \ldots, y_k^k$ are determined iteratively with guaranteed convergence. This iterative scheme was implemented in Caffarelli et al. (1999) in a numerical algorithm for solving Equation (8.25) and concrete numerical examples were presented there. We will not repeat here the details of this algorithm and make only several general comments. In order to make the reference to this algorithm easier and emphasize the fact that it is a special case of the SQM, we will call it the **supporting paraboloid** (SP) algorithm.

The results in Caffarelli et al. (1999) and Kochengin and Oliker (2003) imply that for a fixed k the SP algorithm finds a reflector R^k approximating the solution of the semi-discrete FFR problem with accuracy up to ϵ^k or, more precisely, the determined reflector R^k satisfies

$$e(R^k) \equiv \frac{\sqrt{\sum_{i=1}^k \left(G(R^k, y_i^k) - F_i^k \right)^2}}{\int_D I d\sigma(m)} \le \epsilon^k, \tag{8.28}$$

where ϵ^k is the required error bound. Equation (8.28) is used as a normalized measure of error.

The SP algorithm can be initialized with any reflector $R^k \in \mathcal{R}$. In our implementation the algorithm automatically determines from the data some values of

$p_2^k,\ldots,p_k^k$ for which it is guaranteed that R^k, generated with paraboloids with focal parameters $\left(p_1^k, p_2^k,\ldots,p_k^k\right)$, is in $\mathcal{R}$. Once the algorithm is initialized it produces iteratively a sequence of reflectors R_s^k, $s = 0,1,2,\ldots$, such that (a) each member of this sequence belongs to R, (b) the sequence monotonically converges to a solution of Equation (8.25), and (c) there exists some finite N for which R_N^k satisfies (8.28). This N can be estimated in terms of the given data. More details on the proof of convergence, including an estimate of convergence rate, are in Caffarelli et al. (1999). In general, because of strong nonlinearity of the problem the convergence of the SP algorithm may be slow for large values of k. A rigorous analysis of the SP algorithm in Caffarelli et al. (1999) shows that it generates a sequence of reflectors converging at least linearly. See also de Castro et al. (2016) for related results.

8.9 BRIEF RELATED AND HISTORICAL COMMENTS

The reader may wonder if the SQM limits the class of possible solutions of the FFR problem to solutions that can be represented in terms of supporting paraboloids and limits of sequences of such solutions, similar to the traditional approach limiting the admissible solution to optics with a priori assumed symmetries. This is an important question which does not have a simple answer. Certainly, the surfaces constructed with the SQM are not required to have any symmetries. However, because of the nonlinearities in the partial differential equations describing the problem it is natural to expect several classes of solutions (Oliker, 2008b). In particular, solutions of type B discussed in Section 3.4 form another class of solutions to the FFR problem. Solutions in this class are only piece-wise convex but not globally convex. The difficulties associated with this issue have already been pointed out by B. Kinber (Kinber, 1984). Setting up specific boundary conditions may not always remove the ambiguity and may further complicate the problem. A seemingly reasonable approach is to identify classes of solutions and try to establish uniqueness within such classes. For example, for the FFR problem the class of convex solutions is of particular interest as it may be easier to analyze and fabricate (Malyak, 1992).

Because of important practical applications, the FFR problem has been discussed extensively in engineering literature. Essentially three different approaches have been used to formulate the general problem analytically and solve it numerically. In the first approach the problem is formulated as a system of first order partial differential equations and a method resembling the method of characteristics is applied to solve the system numerically; see Galindo-Israel et al. (1987), Galindo-Israel et al. (1991a), and other references there. In the second approach the same problem is formulated as a boundary value problem for a second order partial differential equation of the Monge-Ampère type for a certain complex-valued function (Westcott, 1983). Then a linearization procedure is used to construct a solution close to some a priori selected solution. In both approaches a rigorous mathematical analysis of the resulting equations was lacking and consequently the validity of the numerics was never fully established. This led to a controversy regarding existence and uniqueness of solutions that has still not been resolved (Kinber, 1984; Galindo-Israel et al., 1987; Galindo-Israel et al., 1991a). The approach presented in previous sections is mathematically rigorous in its formulation, analysis, and numerics. In the next section this approach is further developed and connected with calculus of variations.

REFERENCES

Born, M., and Wolf, E. (1999). *Principles of Optics*, 7th Ed., Section 3.1.2. Cambridge University Press, Cambridge.

Brenner, K.-H. (2010). General solution of two-dimensional-shaping with two surfaces. In *Information Optics and Photonics, Algorithms, Systems, and Applications*, ed. by T. Fournel and B. Javidi, pp. 3–12, Springer, New York.

Burkhard, D. G., and Shealy, D. L. (1987). A different approach to lighting and imaging: Formulas for flux density, exact lens and mirror equations and caustic surfaces in terms of the differential geometry of surfaces. In *SPIE Vol. 692 Materials and Optics for Solar Energy Conversion and Advanced Lighting Technology*, pages 248–272.

Caffarelli, L. A., and Oliker, V. (2008). Weak solutions of one inverse problem in geometric optics. Preprint, 1994. *J. Math. Sci.* **154**(1), 37–46.

Caffarelli, L. A., Gutiérrez, C. E., and Huang, Q. (2008). On the regularity of the reflector antennas. *Ann. Math.* **167**(1), 299–323.

Caffarelli, L. A., Kochengin, S., and Oliker, V. (1999). On the numerical solution of the problem of reflector design with given far-field scattering data. *Contemp. Math.* **226**, 13–32.

de Castro, P. M. M., Mérigot, Q., and Thibert, B. (2016). Far-field reflector problem and intersection of paraboloids. *Numer. Math.* **134**, 389–411.

Fournier, F. R. (2010). *Freeform Reflector Design with Extended Sources* (Ph.D. Diss., University of Central Florida).

Fournier, F. R., Cassarly, W. J., and Rolland, J. P. (2009). Designing freeform reflectors for extended sources. In *SPIE Nonimaging Optics: Efficient Design for Illumination and Solar Concentration VI*, SPIE Vol. 7423, pages 203–214, San Diego, CA.

Galindo-Israel, V., Imbriale, W. A., and Mittra, R. (1987). On the theory of the synthesis of single and dual offset shaped reflector antennas. *IEEE Trans. Antennas Propag.* **AP-35**(8), 887–896.

Galindo-Israel, V., Imbriale, W. A., Mittra, R., and Shogen, K. (1991a). On the theory of the synthesis of offset dual-shaped reflectors – case examples. *IEEE Trans. Antennas Propag.* **39**(5), 620–626.

Galindo-Israel, V., Rengarajan, S., Imbriale, W. A., and Mittra, R. (1991b). Offset dual-shaped reflectors for dual chamber compact ranges. *IEEE Trans. Antennas Propag.* **39**(7), 1007–1013.

Galindo-Israel, V., Mittra, R., and Cha, A. G. (1979). Aperture amplitude and phase control on offset dual reflectors. *IEEE Trans. Antennas Propag.* **AP-27**, 154–164.

Glimm, T., and Oliker, V. (2003). Optical design of single reflector systems and the Monge-Kantorovich mass transfer problem. *J. Math. Sci.* **117**(3), 4096–4108.

Glimm, T., and Oliker, V. (2004). Optical design of two-reflector systems, the Monge-Kantorovich mass transfer problem and Fermat's principle. *Indiana Univ. Math. J.* **53**(5), 1255–1278.

Hasanis, T., and Koutroufiotis, D. (1985). The characteristic mapping of a reflector. *J. Geom.* **24**, 131–167.

Keller, J. B. (1958). The inverse scattering problem in geometrical optics and the design of reflectors. *IEE Transactions on Antennas and Propagation*, **7**(5), 146–149.

Kinber, B. E. (1962). On two reflector antennas. *Radio Eng. Electron. Phys.* **7**(6), 973–979.

Kinber, B. E. (1984). *Inverse Problems of the Reflector Antennas Theory - Geometric Optics Approximation*. preprint No. 38, Academy of Sc., USSR, in Russian.

Kochengin, S., and Oliker, V. (1997). Determination of reflector surfaces from near-field scattering data. *Inverse Probl.* **13**(2), 363–373.

Kochengin, S., and Oliker, V. (1998). Determination of reflector surfaces from near-field scattering data II. Numerical solution. *Numer. Math.* **79**(4), 553–568.

Kochengin, S., and Oliker, V. (2003). Computational algorithms for constructing reflectors. *Comput. Visualization Sci.* **6**, 15–21.

Lieb, E. H., and Loss, M. (1997). *Analysis.* Graduate Studies in Mathematics, v. 14, AMS.

Loeper, G. (2011). Regularity of optimal maps on the sphere: The quadratic cost and the reflector antenna. *Arch. Rational Mech. Anal.* **199**, 269–289.

Malyak, P. H. (1992). Two-mirror unobscured optical system for reshaping irradiance distribution of a laser beam. *Appl. Opt.* **31**(22), 4377–4383.

Mérigot, Q., and Oudset, É. (2016). Discrete optimal transport: Complexity, geometry and applications. *Discrete Compu. Geom.* **55**(2), 263–283.

Michaelis, D., Schreiber, P., and Bruer, A. (2011). Cartesian oval representation of freeform optics in illumination systems. *Opt. Lett.* **36**(6), 918–920.

Newman, E., and Oliker, V. (1994). Differential-geometric methods in design of reflector antennas. In *Symposia Mathematica*, volume 35, pages 205–223.

Oliker, V. (1987). Near radially symmetric solutions of an inverse problem in geometric optics. *Inverse Probl.* **3**, 743–756.

Oliker, V. (1996). The reflector problem for closed surfaces. *Partial Differ. Equ. Appl.* **177**, 265–270.

Oliker, V. (2006a). A rigorous method for synthesis of offset shaped reflector antennas. *Comput. Lett.* **2**(1–2), 29–49.

Oliker, V. (2008a). A Minkowski-style theorem for focal functions of compact convex reflectors. *Trans. Am. Math. Soc.* **360**(2), 563–574.

Oliker, V. (2008b). On design of freeform refractive beam shapers, sensitivity to figure error and convexity of lenses. *J. Opt. Soc. Am. A*, **25**(12), 3067–3076.

Oliker, V. (2011). Designing freeform lenses for intensity and phase control of coherent light with help from geometry and mass transport. *Arch. Rational Mech. Anal.* **201**, 1013–1045.

Oliker, V. (2017). Controlling light with freeform multifocal lens designed with supporting quadric method (SQM). *Opt. Express*, **25**(4), A59–A72.

Oliker, V. I. (2002). On the geometry of convex reflectors. *PDE's, Submanifolds and Affine Differential Geometry, Banach Center Publ.*, **57**, 155–169.

Oliker, V. I. (2008). On the geometry of convex reflectors, II. Polar reflectors. *Results in Math.* **52**, 359–367.

Oliker, V., and Newman, E. (1993). On the energy conservation equation in the reflector mapping problem. *Appl. Math. Lett.* **5**, 91–95.

Oliker, V., and Prussner, L. D. (1994). A new technique for synthesis of offset dual reflector systems. In *10-th Annual Review of Progress in Applied Computational Electromagnetics*, pages 45–52.

Oliker, V., and Waltman, P. (1987). Radially symmetric solutions of a Monge-Ampere equation arising in a reflector mapping problem. In *Proceedings of UAB International Conference on Differential Equations and Mathematical Physics*, ed. by I. Knowles and Y. Saito, pages 361–374. Lecture Notes in Mathematics 1285.

Oliker, V., Prussner, L. D., Shealy, D., and Mirov, S. (1994). Optical design of a two-mirror asymmetrical reshaping system and its application in superbroadband color center lasers. In *SPIE Proceedings*, volume 2263, pages 10–18, San Diego, CA.

Oliker, V., Rubinstein, J., and Wolansky, G. (2015). Supporting quadric method in optical design of freeform lenses for illumination control of a collimated light. *Adv. Appl. Math.* **62**, 160–183.

Guan, P., and Wang, X. J. (1998). On a Monge-Ampère equation arising in geometric optics. *J. Differ. Geom.* **48**, 205–223.

Ries, J., and Muschaweck, J. (2002). Tailored freeform optical surfaces. *J. Opt. Soc. Am. A* **19**(3), 590–595.

Romijn, L. B., ten Thije Boonkkamp, J. H. M., and LJzerman, W. L. (2019). Inverse reflector design for a point source and far-field target. Preprint.

Schneider, R. (2014). *Convex Bodies: The Brunn-Minkowski Theory*, 2nd Expanded Ed. Cambridge University.

Schruben, J. S. (1974). Analysis of rotationally symmetric reflectors for illuminating systems. *J. Opt. Soc. Am.* **64**(1), 55–58.

Taylor, M. E. (2006). *Measure Theory and Integration*. Graduate Studies in Mathematics, v. 76, AMS.

Wang, X.-J. (2004). On design of a reflector antenna II. *Calculus of Variations and PDE's* **20**, 329–341.

Westcott, B. S. (1983). *Shaped Reflector Antenna Design*. Research Studies Press, Letchworth, England.

Westcott, B. S., and Norris, A. P. (1975). Reflector synthesis for generalized far fields. *J. Phys. A: Math. Gen.* **8**, 521–532.

Wu, R., Xu, L., Liu, P., Zhang, Y., Zheng, Z., Li, H., and Liu, X. (2013). Freeform illumination design: A nonlinear boundary problem for the elliptic Monge-Ampére equation. *Opt. Lett.* **38**(2), 229–231.

Wang, X. J. (1996). On design of reflector antenna. *Inverse Probl.* **12**(2), 351–375.

9 Variational Approach

In the far-field problem the reflected *directions* are identified with points on the unit sphere $\mathbb{S}^2$. The target irradiance (often referred to as intensity or irradiance intensity) in this case is the flux per unit area of the aperture on $\mathbb{S}^2$ representing the directions of reflected rays, *W/sr*; see Figure 7.1.

It follows from the discussion in the previous section that the numerical solution of the semi-discrete version of the FFR problem is the critical step. The variational formulation of the FFR problem in the weak form Equation (8.17) described later permits its reduction to a linear programming problem. Furthermore, the semi-discrete version upon a suitable discretization of the source radiance becomes a finite-dimensional linear programming (LP) problem whose solution is an approximation to the solution of the original problem. The LP setting for the FFR problem can be put in a form acceptable to available codes.

It is important to recognize that unlike the SQM the LP formulation **is not possible** in all optical design problems and in each case the required optimization problem must be formulated anew. In some cases, for example in near-field reflector/refractor problems with one mirror/lens, a linear programming formulation is not available.

The organization of this section is as follows. First, in Section 9.1 we describe a special linear functional (9.5) and state a theorem establishing equivalence between solutions of (8.17) and maximizers of (9.5). In Section 9.2 we specialize the functional (9.5) to the semi-discrete case. In Section 9.3 we present a special discretization of the source radiance and formulate the corresponding LP problem. Finally, in Section 9.6 we present important general remarks indicating limitations on the applicability of the variational approach.

Everywhere in this section we continue to use the notations introduced in previous sections. The presentation here is a slightly modified version of presentations in our papers Oliker (2005) and Glimm and Oliker (2003).

9.1 THE FFR PROBLEM AS AN OPTIMIZATION PROBLEM

We continue to denote by $\bar{D}, \bar{T} \subset \mathbb{S}^2$ two closed domains in $\mathbb{S}^2$ (the input and output apertures; see Fig. 7.1). As before, $I(m)$ and $L(y)$ are nonnegative integrable functions given on $\bar{D}$ and $\bar{T}$, respectively, satisfying Equation (8.21), and extended to the entire $\mathbb{S}^2$ as zero in $\mathbb{S}^2 \setminus \bar{D}$ and $\mathbb{S}^2 \setminus \bar{T}$, respectively. Again, m denotes a point in $\bar{D}$ and $\mathbb{S}^2 \setminus \bar{D}$ and y in $\bar{T}$ and $\mathbb{S}^2 \setminus \bar{T}$. The set of closed convex reflectors as in Definition 3.1 is denoted, as before, by $\mathcal{R}$. We also use the same notation p to denote the extended focal function of $R \in \mathcal{R}$.

For $m, y \in \mathbb{S}^2$ put

$$\mathcal{K}(m,y) = \begin{cases} -\log(1 - \langle m, y \rangle) & \text{if } m \neq y, \\ +\infty & \text{if } m = y. \end{cases} \tag{9.1}$$

It follows from (8.13)–(8.14) that any pair of functions $\rho, p \in C(\mathbb{S}^2)$ such that $\rho, p > 0$ defines uniquely a reflector $R \in \mathcal{R}$ if

$$\log \rho(m) - \log p(y) \leq \mathcal{K}(m,y) \forall m, y \in \mathbb{S}^2 \tag{9.2}$$

and for each m there is a y and for each y there is an m such that

$$\log \rho(m) - \log p(y) = \mathcal{K}(m,y). \tag{9.3}$$

The generalized reflector map γ is defined as before in Equation (8.15):

$$\gamma(m) = \left\{ y \in \mathbb{S}^2 \mid \log \rho(m) - \log p(y) = \mathcal{K}(m,y) \right\}, \quad m \in \mathbb{S}^2.$$

Because $\rho, p > 0$ on $\mathbb{S}^2$, bounded, and $\mathcal{K}(m,y) = +\infty$ iff $m = y$, the map γ defined by R has no fixed points and $\mathcal{K}(m,y)$ is bounded whenever (9.3) holds.

We formulate now the maximization problem associated with the FFR problem. First, we define the set of admissible pairs of functions:

$$\text{Adm}_-(\mathbb{S}^2) = \left\{ s, q \in C(\mathbb{S}^2) \mid s(m) - q(y) \leq \mathcal{K}(m,y) \quad \forall m, y \in \mathbb{S}^2 \right\}. \tag{9.4}$$

Evidently, if ρ, p are the radial and focal functions of $R \in \mathcal{R}$ then $(\log \rho, \log p)$ is in $\text{Adm}_-(\mathbb{S}^2)$.

Let I, L be two functions as above. Consider the linear functional

$$\mathcal{F}(s,q) = \int_{\mathbb{S}^2} s(m) I(m) d\sigma(m) - \int_{\mathbb{S}^2} q(y) L(y) d\sigma(y), \quad s, q \in C(\mathbb{S}^2) \tag{9.5}$$

and the maximization problem

$$\mathcal{F}(s,q) \to \max \text{ over } (s,q) \in \text{Adm}_-(\mathbb{S}^2). \tag{9.6}$$

Clearly, $\mathcal{F}$ is continuous on $C(\mathbb{S}^2) \times C(\mathbb{S}^2)$ relative to the norm defined as $\max \left\{ \max_{\mathbb{S}^2} |s|, \max_{\mathbb{S}^2} |q| \right\}$.

Theorem 9.1.1 *Let $R \in \mathcal{R}$ and $(\rho, p) \in C(\mathbb{S}^2) \times C(\mathbb{S}^2)$ be the corresponding radial and focal functions of R. The following two statements are equivalent:*

(1) *The pair $(\log \rho, \log p)$ is a solution of problem (9.6);*
(2) *The reflector $R = (\rho, p)$ is a weak solution of the FFR problem 8.17.*

An immediate consequence of Theorem 9.1.1 is that the weak solution whose existence is guaranteed by Theorem 8.7.1 can be determined by finding a maximizer of functional $\mathcal{F}$ on $\mathrm{Adm}_-(\mathbb{S}^2)$. The discrete version of this maximization problem can be solved by numerical methods of linear optimization; see Section 9.2. It can be shown that from the point of view of optimization the maximization problem (9.6) is the dual of the minimization problem which will be discussed in Section 9.3. In a slightly different form, Theorem 9.1.1 was proved in Glimm and Oliker (2003). The connection of the FFR problem with optimization was also independently shown in Wang (2004). Earlier, such connection for the two-mirror reflector problem was established in our paper Glimm and Oliker (2004).

Under an additional geometric assumption that $\bar{D} \cap \bar{T} = \varnothing$ existence of a maximizer of $\mathcal{F}$ can be proved independently of Theorem 8.7.1; see the remark at the end of Section 5 in Glimm and Oliker (2003). It was shown in Gangbo and Oliker (2007) that this result holds under much weaker assumptions which allow the cost function (and their generalizations) to be infinite (Gangbo & Oliker, 2007).

9.2   DISCRETE MAXIMIZATION PROBLEM

In order to solve the maximization problem (9.6) numerically, we need to discretize the functional (9.5) and the constraints in Equation (9.4) and then solve the obtained discrete maximization problem. Below we describe this solution in detail.

Let $M, N \in \mathbb{N}, M \geq 1, N \geq 2$, fix M points $m_1, \ldots, m_M \in \mathbb{S}^2$ and N points $y_1, \ldots, y_N \in \mathbb{S}^2$ and in each case repeat the steps in Section 8.7.1. As a result we get M measurable subsets $\Omega_i^M \subset \mathbb{S}^2, i = 1, \ldots, M$, and N measurable subsets $\omega_j^N \subset \mathbb{S}^2, j = 1, \ldots, N$, such that for each $i = 1, \ldots, M$ diameters $\Omega_i^M < 3\pi / M$ and for each $j = 1, \ldots, N$ diameters $\omega_j^N < 3\pi / N$, and for $1 \leq t, k \leq M, 1 \leq h, l \leq N$,

$$\bigcup_{i=1}^{M} \bar{\Omega}_i^M = \mathbb{S}^2, \Omega_t^M \cap \Omega_k^M = \varnothing \text{ if } t \neq k;$$

$$\bigcup_{j=1}^{N} \bar{\omega}_j^N = \mathbb{S}^2, \omega_h^N \cap \omega_l^N = \varnothing \text{ if } h \neq l.$$

For $i = 1, 2, \ldots, M$ and $j = 1, 2, \ldots, N$ pick points $m_i^M \in \Omega_i^M$ and $y_j^N \in \omega_j^N$, and put

$$E_i^M = \int_{\Omega_i^M} I(m) d\sigma(m), \quad F_i^N = \int_{\omega_j^N} L(y) d\sigma(y).$$

For measurable sets $\Omega, \omega \subset \mathbb{S}^2$ we define

$$E^M(\Omega) = \sum_{m_i^M \in \Omega} E_i^M \text{ and } F^N(\omega) = \sum_{y_i^N \in \omega} F_i^k.$$

Assume that

$$\sum_{i=1}^{M} E_i^M = \sum_{j=1}^{N} F_j^N.$$

Put

$$\text{Adm}_{M,N} = \left\{ s_i^M - q_j^N \le \mathcal{K}\left(m_i, y_j\right), \; m_i \ne y_j, \; \forall i = 1, \ldots, M, \; \forall j = 1, \ldots, N \right\} \quad (9.7)$$

and let

$$\mathcal{F}_{M,N}\left(s_1^M, \ldots, s_M^M, q_1^N, \ldots, q_N^N\right) := \sum_{i=1}^{M} s_i^M E_i^M - \sum_{j=1}^{N} q_j^N F_j^N.$$

The maximization problem is to find $\overline{s}^M = \left(\overline{s}_1^M, \ldots, \overline{s}_M^M\right), \overline{q}^M = \left(\overline{q}_1^N, \ldots, \overline{q}_N^N\right)$ satisfying (9.7) and such that

$$\mathcal{F}_k\left(\overline{s}^M, \overline{q}^N\right) = \sup_{\text{Adm}_{M,N}} \mathcal{F}_{M,N}\left(s^M, q^N\right).$$

Applying Theorem 9.1.1 (in its discretized version) we conclude that this problem is equivalent to the fully discrete problem obtained from the semi-discrete FFR problem (8.25) described in Section 8.7. Specifically, the required additional discretization is that of the apperture $\overline{D}$ (or $\mathbb{S}^2$), which has to be done anyway in problem (8.25) in order to compute $G(R, \omega)$. Thus, the discrete FFR problem is now reduced to a solution of a discrete linear programming (LP) problem which can be done by known numerical methods (Rúshendorf & Uckelmann, 2000); see also Mérigot and Oudset (2016) for related results.

9.3 CONNECTION WITH OPTIMAL MASS TRANSPORT

According to the duality theory in optimization it is possible to construct a minimization problem dual to the maximization problem (9.6). By Theorem 9.1.1 this minimization problem must be related to the FFR problem. We describe now this connection following Section 4 of Glimm and Oliker (2003).

A variant of the celebrated Monge–Kantorovich optimal mass transport problem (Villani, 2009) in Euclidean space may be stated in our situation on $\mathbb{S}^2$ as follows. Let $\mathcal{K}(m, y)$ be as in (9.1). Denote by $\mathcal{P}$ the set of maps $P : \mathbb{S}^2 \to \mathbb{S}^2$ which are "onto" and measure-preserving; that is, they satisfy the substitution rule

$$\int_{\mathbb{S}^2} h\left(P(m)\right) I(m) d\sigma(m) = \int_{\mathbb{S}^2} h(y) L(y) d\sigma(y)$$

for all continuous functions h on $\mathbb{S}^2$. Each such map is called a *plan*. Note that a plan needs only be defined almost everywhere on *sptI*. Here,

$$sptI = \overline{\{m \in \mathbb{S}^2 \mid I(m) \ne 0\}},$$

where the overline indicates the closure of the set.

The Monge–Kantorovich mass transport problem we are interested in is: *among all plans in $\mathcal{P}$ find a plan P minimizing the transportation cost*

$$P \mapsto \int_{\mathbb{S}^2} \mathcal{K}\big(m, P(m)\big) I(m) d\sigma(m). \tag{9.8}$$

(The transportation cost may be infinite for certain plans; for example, the transportation cost is $+\infty$ if the plan is the identity map of $\mathbb{S}^2$ to itself.)

For weak solutions of the reflector problem the generalized reflector map is a plan by Equation (8.19). In fact, we have the following:

Theorem 9.3.1 *Let $R \in \mathcal{R}$, and ρ, p are the corresponding radial and focal functions of R. Suppose R is a solution of the FFR problem in weak form (8.17) and γ its generalized reflector map. Then γ minimizes the transportation cost (9.8) among all plans in $\mathcal{P}$, and any other minimizer is equal to γ almost everywhere on spt(I).*

An important physical conclusion of this theorem is that the cost $\mathcal{K}(m, y)$ encodes all the optical information needed for defining the required reflector. Theorem 3.1 can also be used to prove uniqueness of weak solutions to the FFR problem. The proof of this theorem is in Glimm and Oliker (2003); see Theorem 4.1.

Theorem 9.3.2 (see Glimm & Oliker, 2003) *Let R, $R' \in \mathcal{R}$ be two weak solutions of the FFR problem with the same data. Let ρ, p' be their respective radial functions. Then R and R' differ only by a homothetic transformation; that is, there exists some $c = const > 0$ such that $\rho(m) = c\rho'(m)$ for all $m \in \mathbb{S}^2$.*

Uniqueness was already discussed after Theorem 8.7.1. Theorem 9.3.2 provides an alternative proof based on coincidence of generalized reflector maps.

9.4 COMPUTED REFLECTOR FOR FFR PROBLEM—POINT SOURCE

Here we provide details on the designed with the SQM reflector in Figure 7.2 solving a FFR problem. As before, $\mathbb{S}^2$ is a unit sphere centered at the origin of the Cartesian coordinate system in 3D space $\mathbb{R}^3$.

Point source. The input aperture $\bar{\Omega}$ on $\mathbb{S}^2$, defined by the radiating point source at the center of $\mathbb{S}^2$ (see Figure 9.1), and its radiance intensity are:

$$\bar{\Omega} = \left\{ \frac{\pi}{4} \le \theta \le \frac{3\pi}{4} ; -\frac{\pi}{2} \le \phi \le \frac{\pi}{2} \right\},$$

$$\mathbf{m} = (\theta, \phi),$$

$$I(\theta, \phi) = \begin{cases} \sin\theta & (\theta, \phi) \in \bar{\Omega} \\ 0 & (\theta, \phi) \notin \bar{\Omega} \end{cases}$$

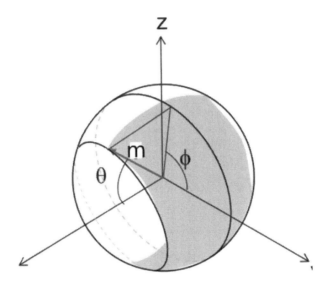

FIGURE 9.1   The shaded region on the unit sphere is the aperture $\overline{\Omega}$ through which the light rays emitted by the point source at the center of the sphere travel into space; **m** is a unit vector in direction of such a ray.

The far-field $\overline{T}$ and output intensity L are:

$$\overline{T} = \left\{ \frac{4\pi}{9} \le \beta \le \frac{5\pi}{9}, \frac{\pi}{6} \le \alpha \le \frac{\pi}{6} \right\},$$

$$y = (\alpha, \beta),$$

$$L(\alpha, \beta) = \begin{cases} \dfrac{k}{\cos^3\alpha\sin^3\beta} & (\alpha, \beta) \in \overline{T} \\ 0 & (\alpha, \beta) \notin \overline{T} \end{cases}$$

The output intensity L is chosen so that the designed reflector produces constant irradiance on a distant plane. The freeform reflector producing the required transformation has already been shown in Figure 7.2.

   Note. Figures 9.1, 9.2, and 9.3 are modified versions of figures from Benítez et al. (2006); reproduced with permission.

9.5   TYPE B REFLECTORS

In this section we mainly follow our paper (Glimm & Oliker, 2003). For ease of reference we will refer to reflectors considered in preceding sections as **reflectors of type A.** An important property of reflectors of type A is that they are convex surfaces. We show now that there is another class of reflectors, called **type B**, that also solve the FFR problem. In contrast to type A reflectors which are guaranteed to be

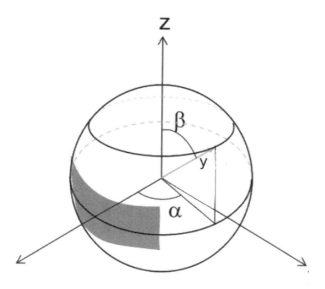

FIGURE 9.2   The far-field $\overline{T}$ is the shaded area; k is a constant enforcing (8.21).

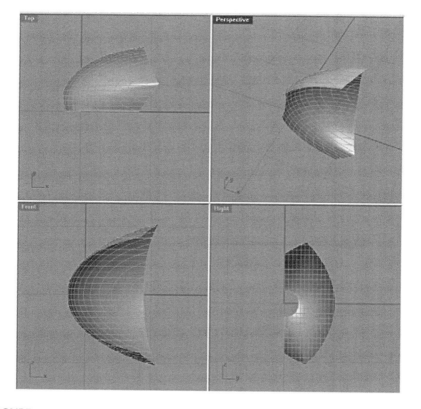

FIGURE 9.3   Side, bottom, and front views of the designed freeform reector.

convex, reflectors of type B are not necessarily convex. This may be a serious disadvantage for manufacturing. Physically, the reflectors of type A produce reflected rays which are crossing while for reflectors of type B the reflected rays are noncrossing. The principal difference in constructions of type B and type A reflectors is that to build a type B reflector one must take a *union* in Definition (8.8).

Definition 9.5.1 *Let $D, T \subset \mathbb{S}^2$ and let (ρ, p) be a pair of functions defined on $\bar{D}$ and $\bar{T}$, respectively. We say that the pair (ρ, p) defines a reflector of type B if $\rho, p > 0$, and*

$$\rho(m) = \sup_{y \in \bar{T}} \frac{p(y)}{1 - \langle m, y \rangle}, \quad m \in \bar{D}, \tag{9.9}$$

$$p(y) = \inf_{m \in \bar{D}} \left[\rho(m) \left(1 - \langle m, y \rangle \right) \right], \quad y \in \bar{T}. \tag{9.10}$$

The surface given by the vector function $r(m) = (m, \rho(m))$, $m \in \bar{D}$ is called a reflector of type B. The set of reflectors of type B is denoted by $\mathcal{R}_{\mathcal{B}}$. See Fig. 9.4 for an illustration in plane.

Figure 9.4: here, a type B reflector consists of two parts: the curve ABC and curve EFG; $\bar{D}$ is a circle of unit radius with center O—the light source, and $\bar{T} = \{y_1, y_2\}$. Both parts are composed of pieces of parabolas $P(y_1)$ and $P(y_2)$. The segments shown by interrupted lines are not part of the reflector. The rays incident on AB and EF

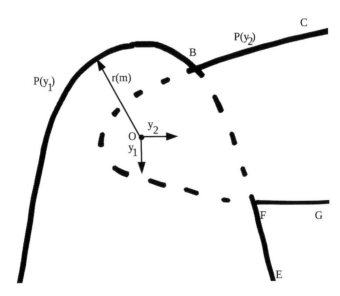

FIGURE 9.4 Here, a type B reflector consists of two parts: the curve ABC and curve EFG; $\bar{D}$ is a circle of unit radius with center O - the light source, and $\bar{T} = \{y_1, y_2\}$. Both parts are composed of pieces of parabolas $P(y_1)$ and $P(y_2)$. The segments shown by interrupted lines are not part of the reflector. The rays incident on AB and EF are reflected in direction y_1 and the rays incident on BC and FG are reflected in direction y_2.

are reflected in direction y_1 and the rays incident on BC and FG are reflected in direction y_2.

Condition (9.9) means that in order to define a type B reflector, for each $m \in \bar{D}$, one has to choose among paraboloids with axes $y \in \bar{T}$ and focal parameters $p(y)$, $y \in \bar{T}$, the most distant from point O in direction m. On the other hand, condition (9.10) means that for each $y \in \bar{T}$ one has to choose the paraboloid with axis y and focal parameter realizing the shortest distance from O to the directrix plane of that paraboloid. Thus, the first condition is geometric while the second is optical and they must be consistent to define a reflector of type B. A similar interpretation can be given to conditions (8.13)–(8.14) taking into account that inf and sup are interchanged as compared to Equations (9.9)–(9.10).

Type B reflectors can be constructed geometrically as follows. Consider the paraboloids of revolution $P(y)$, $y \in \bar{T}$, with focal functions $p(y)$ and the convex bodies $B(y)$ bounded by $P(y)$. Put

$$B = \bigcup_{y \in \bar{T}} B(y) \text{ and } R = (\partial B) \cap K(\bar{D}),$$

where $K(\bar{D})$ is the cone of rays emanating from O and passing through points of $\bar{D}$. Clearly, a reflector $R \in \mathcal{R}_B$ and it is a graph of ρ over $\bar{D}$. With the above definition it is guaranteed that for each $y \in \bar{T}$ the paraboloid given by $\rho_y(m) = p(y)/(1 - \langle m, y \rangle)$, $m \in \bar{D}$ has a point of contact with R. Note that the concept of supporting paraboloid does not generalize to reflectors of type B.

For our purposes it will be convenient to rewrite the characterization of type B reflectors in the following equivalent form:

$$\log \rho(m) - \log p(y) \geq \mathcal{K}(m, y) \quad \forall m \in \bar{D}, \ \forall y \in \bar{T}, \tag{9.11}$$

and where for each m there is a y such that equality is achieved and vice versa.

Assume now that the sets $D, T \subset \mathbb{S}^2$ are both nonempty, open and connected, and satisfy the condition

$$\bar{D} \cap \bar{T} = \varnothing. \tag{9.12}$$

The results below remain true for more general subsets of $\mathbb{S}^2$—as long as they satisfy (9.13). Because of the geometric restriction $\bar{D} \cap \bar{T} = \varnothing$ we have the following:

Lemma 5.2 *Suppose D and T satisfy condition (9.12) and let (ρ, p) be a pair of function* $(\rho, p) \in C(\bar{D}) \times C(\bar{T})$ *defining type A or B reflectors. Put* $C_1 = \min_{(m,y) \in \bar{D} \times \bar{T}} d(m, y)$. *Then for any $m, m' \in \bar{D}$*

$$\left| \log \rho(m') - \log \rho(m) \right| \leq \frac{(4 + C_1)\pi}{C_1} \sqrt{\frac{2}{1 - \cos(C_1/2)}} d(m', m). \tag{9.13}$$

Also, for any $y, y' \in \bar{T}$

$$\left| \log p(y') - \log p(y) \right| \leq \frac{(4 + C_1)\pi}{C_1} \sqrt{\frac{2}{1 - \cos(C_1/2)}} d(y', y). \tag{9.14}$$

See the proof in Glimm and Oliker (2003).

Functions satisfying conditions (9.13) and (9.14) are called Lipschitz continuous. It is known that Lipschitz continuous functions are differentiable almost everywhere in D and T (cf. Notes 8.5.1 and 8.5.2 at the end of Section 8.5). Thus, such are the radial and focal functions, ρ and p, of any reflector of type B (or A), assuming a priori that $(\rho, p) \in C(\bar{D}) \times C(\bar{T})$.

Consequently, for type B reflectors, similar to the case of reflectors of type A, we can use (9.11) to define the generalized ray-tracing map associated with R; namely, we put

$$\gamma(m) = \left\{ y \in \bar{T} \mid \log \rho(m) - \log p(y) = \mathcal{K}(m, y) \right\}, \quad m \in \bar{D}.$$

It follows from differentiability almost everywhere of ρ and p that $\gamma(m)$ is single-valued almost everywhere in D.

It follows from Equation (9.11) that the map γ is onto, that is, $\gamma(\bar{D}) = \bar{T}$. In addition,

$$G(R, \omega) = \int_{\gamma^{-1}(\omega)} I(m) d\sigma(m) \quad \text{for any measurable set } \omega \subseteq T.$$

is a measure on T for any nonnegative and integrable function I on D.

Suppose we are given two nonnegative and integrable functions I and L on D and T, respectively, satisfying the conservation law in (8.24). The far-field reflector problem for reflectors in $\mathcal{R}_B$ is to find $R \in \mathcal{R}_B$ such that

$$G(R, \omega) = \int_{\omega} L(m) d\sigma(m) \text{ for any measurable set } \omega \subseteq T.$$

Now we describe the variational formulation of the FFR problem for reflectors of type B. The constraints on the admissible functions are

$$\text{Adm}_+ \left(\bar{D}, \bar{T} \right) = \left\{ (s, q) \in C(\bar{D}) \times C(\bar{T}) \text{ and} \right.$$

$$\left. s(m) - q(y) \geq \mathcal{K}(m, y) \text{ for all } m \in \bar{D}, y \in \bar{T}. \right\}$$

Theorem 9.5.3 *(See Glimm & Oliker, 2004.) Suppose $\bar{D}$ and $\bar{T}$ satisfy (9.12) and let $R \in \mathcal{R}_B$ be defined by a pair $(\rho, p) \in C(\bar{D}) \times C(\bar{T})$. Put*

$$\mathcal{F}(s, q) = \int_D s(m) I(m) d\sigma(m) - \int_{\bar{T}} q(y) L(y) d\sigma(y), s, q \in C(\mathbb{S}^2),$$

where I and L satisfy

$$\int_{\bar{\Omega}} I(m) d\sigma(m) = \int_{\bar{T}} L(y) d\sigma(y) \neq 0.$$

The following assertions are equivalent:

(1) $\min_{\text{Adm}_+(\bar{D}, \bar{T})} \mathcal{F}(s, q) = \mathcal{F}(\rho, p).$

(2) *The reflector R defined by (ρ, p) is a weak solution of the FFR problem.*

For the Monge–Kantorovich problem (9.8) we have the following:

Theorem 9.5.4 *Suppose $\bar{D}$ and $\bar{T}$ satisfy condition (9.12) and let $R \in \mathcal{R}_\mathcal{B}$ be a weak solution of the FFR problem. Let γ be its ray-tracing map. Then γ maximizes the transportation cost* $P \mapsto \int_D \mathcal{K}(m, P(m)) I(m) \, d\sigma(m)$ *among all plans* $P : \bar{D} \to \bar{T}$, *and any other maximizer is equal to γ almost everywhere on* $spt(I) \setminus \{m \in D \mid I(m) = 0\}$.

The existence of type B solutions to the reflector problem is established by showing that $\mathcal{F}$ admits a minimum on $\mathrm{Adm}_+(D,T)$. See Glimm and Oliker (2004) for details and a uniqueness statement. Here, we just note that under condition (9.12) the existence of weak solution is proved independently in Theorem (8.7.1).

9.6   WHEN DOES THE VARIATIONAL APPROACH NOT APPLY?

In the FFR problem the cost function (9.1) was defined as

$$\mathcal{K}(m, y) = -\log(1 - \langle m, y \rangle), m, y \in \mathbb{S}^2.$$

The radial and focal functions ρ and p are connected with $\mathcal{K}$ by Equation (9.2). However, in many optical problems the focal function enters into the expression for the cost. Consequently, the resulting optimization problem is not linear and the known methods can not be directly applied for solving the problem. For example, in the near-field reflector (NFR) problem, one is given a point source of light and a target set at a finite distance. It is required to determine a reflector that intercepts the light rays emitted by the source and reflects them so that a given irradiance distribution is produced on the target. This problem can be solved for the weak solution with the SQM (Kochengin & Oliker, 1997). However, this weak solution is not an optimizer of the functional associated with the NFR problem. An explicit example illustrating this situation is provided in Graf and Oliker (2012).

9.7   STRONG SOLUTIONS OF THE FFR PROBLEM

In this section we discuss only the reflectors of type A.

It was shown in Oliker (2002) that if a given function $p \in C^2(\mathbb{S}^2)$ is such that $p > 0$ on $\mathbb{S}^2$ and

$$p_{\alpha\alpha} + p - \rho > 0,$$

where $p_{\alpha\alpha}$ denotes differentiation along the arc length of any large circle of $\mathbb{S}^2$ and $\rho = \dfrac{|\nabla p|^2 + p^2}{2p}$ then the map

$$-r(y) = \nabla p(y) + p(y)y - \rho(y)y : \mathbb{S}^2 \to R^3, \tag{9.15}$$

defines a reflector in $\mathcal{R}$ of class C^1. Furthermore, $|r(y)| = \rho(y)$; that is, the value of the radial function in direction $m(y) = |r(y)| / \rho(y)$ is $\rho(y)$.

The question when a positive continuous function $\rho(y)$, $y \in \mathbb{S}^2$, generates a reflector in $\mathcal{R}$ such that for each $y \in \mathbb{S}^2$ the paraboloid

$$P(y): \rho(m) = \frac{p(y)}{1-\langle m, y \rangle}$$

is supporting was investigated in Oliker (2008).

If the focal function p of a reflector $R \in \mathcal{R}$ is positive and of class $C^2(T) \cap C^1(\bar{T})$ then the Jacobian of the inverse map $\gamma^{-1}: \bar{T} \to \bar{D}$ in Equation (8.18) can be computed as

$$M(p) \equiv \left| J\left(\gamma^{-1}\right) \right| = \frac{det\left[Hess(p) + (p-\rho)e \right]}{\rho^2 det(e)},$$

where *Hess* (p) is the matrix of second covariant derivatives; see Westcott and Norris (1975) and Oliker and Waltman (1987). Then, for a measurable set $\omega \subset T$

$$G(R,\omega) = \int_{V(\omega)} I(m)d\sigma(m) = \int_{\omega} I\left(r(y)/\rho(y)\right) M\left(p(y)\right) d\sigma(y).$$

The point-wise Equation (8.18) assumes now the form

$$I\left(\gamma^{-1}(y)\right) M\left(p(y)\right) = L(y) \text{ on } T \subset \mathbb{S}^2. \tag{9.16}$$

The reflector problem in this case is formulated as the problem of finding a function $p \in C^2(T) \cap C^1(\bar{T})$ such that the normalized map (9.15)

$$r/\rho : \bar{T} \to \bar{D}$$

is one-to-one between $\bar{D}$ and $\bar{T}$ and satisfies (9.16). Such a function p is called a *strong* solution of the reflector problem. Regularity (that is, differentiability of weak solutions) under various assumptions on I and L was studied in Xu-Jia Wang (1996); Guan and Wang (1998); Caffarelli et al. (2008); and Loeper (2011).

9.8 ACKNOWLEDGMENT

This work presented in Chapters 7–9 is supported in part by the U.S. Air Force Office of Scientific Research (AFOSR) under award FA9550-18-1-0189.

REFERENCES

Benítez, P., Winston, R., and Miñano, J. C. (2006). *Nonimaging Optics*. Short Course at SPIE annual meeting, San Diego, CA, # SC388. 76–78.
Caffarelli, L. A., Gutiérrez, C. E., and Huang, Q. (2008). On the regularity of the reflector antennas. *Ann. Math.* 167(1), 299–323.
Gangbo, W., and Oliker, V. (2007). Existence of optimal maps in the reflector-type problems. *ESAIM: Control, Optim. Calculus Var.* 13(1), 93–106.

Glimm, T., and Oliker, V. (2003). Optical design of single reflector systems and the Monge-Kantorovich mass transfer problem. *J. Math. Sci.* **117**(3), 4096–4108.

Glimm, T., and Oliker, V. (2004). Optical design of two-reflector systems, the Monge-Kantorovich mass transfer problem and Fermat's principle. *Indiana Univ. Math. J.* **53**(5), 1255–1278.

Graf, T., and Oliker, V. I. (2012). An optimal mass transport approach to the near-field reflector problem in optical design. *Inverse Probl.* **28**(2), 1–15.

Guan, P., and Wang, X. J. (1998). On a Monge-Ampère equation arising in geometric optics. *J. Differ. Geom.* **48**, 205–223.

Kochengin, S., and Oliker, V. (1997). Determination of reflector surfaces from near-field scattering data. *Inverse Probl.* **13**(2), 363–373.

Loeper, G. (2011). Regularity of optimal maps on the sphere: The quadratic cost and the reflector antenna. *Arch. Rational Mech. Anal.* **199**, 269–289.

Mérigot, Q., and Oudset, É. (2016). Discrete optimal transport: Complexity, geometry and applications. *Discrete Compu. Geom.* **55**(2), 263–283.

Oliker, V. (2005). Geometric and variational methods in optical design of reflecting surfaces with prescribed irradiance properties. In *SPIE Proceedings, Nonimaging Optics and Efficient Illumination Systems II*, ed. by R. Winston and R. J. Koshel, volume 5942, pages 07-1–07-12, San Diego, CA.

Oliker, V. I. (2002). On the geometry of convex reflectors. *PDE's, Submanifolds and Affine Differential Geometry, Banach Center Publ.*, **57**, 155–169.

Oliker, V. I. (2008). On the geometry of convex reflectors, II. Polar reflectors. *Results in Math.* **52**, 359–367.

Oliker, V., and Waltman, P. (1987). Radially symmetric solutions of a Monge-Ampere equation arising in a reflector mapping problem. In *Proceedings of UAB International Conference on Differential Equations and Mathematical Physics*, ed. by I. Knowles and Y. Saito, pages 361–374. Lecture Notes in Mathematics 1285.

Rúshendorf, L., and Uckelmann, L. (2000). Numerical and analytical results for the transportation problem of Monge-Kantorovich. *Metrika* **51**(3), 245–258 (electronic).

Villani, C. (2009). *Optimal Transport: Old and New*. Grundlehren der Mathematischen Wissenschaften, Volume 338, Springer-Verlag, Berlin.

Wang, X. J. (1996). On design of reflector antenna. *Inverse Probl.* **12**(2), 351–375.

Wang, X.-J. (2004). On design of a reflector antenna II. *Calculus of Variations and PDE's* **20**, 329–341.

Westcott, B. S., and Norris, A. P. (1975). Reflector synthesis for generalized far fields. *J. Phys. A: Math. Gen.* **8**, 521–532.

10 A Paradigm for a Wave Description of Optical Measurements

10.1 INTRODUCTION

Radiance, which is the density of radiative power in phase space, has been the subject of a rich literature over the past 40 years and more (Littlejohn & Winston, 1993; Walther, 1973; Wolf, 1978). It is a tribute to the work of Adrian Walther, Emil Wolf, and many others that the development of a wave theory of radiance, known as "generalized radiance," continues to the present time. This chapter is complementary to this line of development in that we attempt to bridge the gap between theory and practical radiometry. Radiometric measurements are important in many branches of science and technology. For example, in illumination engineering, the visibility of displays is quantified by radiometers. In astrophysics, radiometry in the far-infrared has played a critical role in understanding the large space-time structure of the universe (Mather, Toral, & Hemmati, 1986). Of course, in the short wave length limit where diffraction effects can be neglected, geometrical optics suffices and one can dispense with the technical difficulties that the wave property of light introduces. But it is precisely in the regime where diffraction effects cannot be neglected that the properties of the measuring instrument have to be taken into account. Moreover, in the absence of a consistent formalism which does take the diffraction property of the instrument into account, it may be difficult to assess the significance of such effects. "Back of the envelope" estimates of diffraction effects may not be reliable and the practical radiometer is left with little guidance as to the magnitude of such effects. For example, an excellent text on radiometry famously states that diffraction effects are "beyond the scope" of the book. In papers (Sun et al., 2002; Winston, Sun, & Littlejohn, 2002) we have shown how the measurement of radiance can be understood in terms of the statistical properties of the electromagnetic field and the properties of the instrument. However, the utility of this approach was limited by the availability of accessible instrument functions that represent the measuring apparatus. In the process we exhibited a remarkable analogy between the result of measuring radiance and the van Cittert–Zernike (VCZ) theorem. In this chapter we first give an overview of a wave description of the measurement of radiance, referring details to previous publications. This approach is validated by experimental results with highly sophisticated radiometers. The excellent agreement with the analytical model suggests that while our demonstration was confined to the measurement of radiance, it is likely that similar considerations apply to a wide class of optical measurements, where diffraction effects are significant.

FIGURE 10.1 Nonimaging collectors in the Sudbury neutrino detector in the Sudbury Neutrino Observatory.

We have seen that the concept of étendue plays a central role in nonimaging optics. It is therefore natural to give étendue a wave theoretic description, since light is a wave phenomenon. The closest connection to traditional optics is the concept of radiance and its measurement. Radiance (or its twin relative "brightness") is extremely useful in illumination, signal detection, and the like. It is therefore natural to seek a wave basis for its description. In fact, as we will see, diffraction effects can be significant in certain regimes, and need to be properly accounted for in a complete and correct theory of optics. Moreover, radiometry (the science of measuring radiance) is not just a relic of the past, but is very much at the cutting edge of current topics. An example are the 10,000 nonimaging cones (Figure 1) used to collect Cherenkov light in the Sudbury neutrino detector, a recipient of the most recent physics Nobel prize. Figure 10.1 shows a light collecting cone from the COBE far-infrared radiometer (Nobel prize 2006).

10.2   THE VAN CITTERT–ZERNIKE THEOREM

The well-known van Cittert–Zernike theorem states that for an incoherent, quasimonochromatic source of radiation, the equal-time degree of coherence (two-point correlation function) $\Gamma\left(\vec{r},\vec{r}'\right)$ is proportional to the complex amplitude in a certain diffraction pattern: the amplitude at $\vec{r}$ is formed by a spherical wave converging to $\vec{r}'$ and diffracted by an aperture the same size, shape, and location as the source (Born & Wolf, 1999; Mandel & Wolf, 1995). The source could, for example, be a thermal blackbody followed by a filter that selects a small wavelength range.

A familiar geometry is a circular source. Then, apart from a normalizing factor, $\Gamma(\vec{r},\vec{r}')$ in a transverse plane becomes the well-known Airy diffraction amplitude, Equation (10.1).

$$\Gamma\left(\vec{r},\vec{r}'\right) = \left(\text{const}\right) F\left(ks\theta_s\right) \qquad (10.1)$$

where $F(x) = 2J_1(x)/x, k = 2\pi/\lambda$, θ_s is the angle subtended by the source at $\vec{r}$ or $\vec{r}'$, and where $s = |\vec{r} - \vec{r}'|$. Recall that $\Gamma(\vec{r}, \vec{r}')$ has its first zero at $s_1 = 0.61\lambda/\theta_s$. For a numerical example, we consider terrestrial sunlight. Then θ_s is 4.7mrad, so that for $\lambda = 0.5$ mm, s_1 is approximately 65 mm. This is the scale of the transverse correlation of sunlight.

10.3 MEASURING RADIANCE

In a previous paper (Littlejohn & Winston, 1995), we examined the relationship between the generalized radiance and the measuring process. We showed how this process can be quantified by introducing the *instrument function*, which is a property of the measuring apparatus (Sun et al., 2002; Winston, Sun, & Littlejohn, 2002). We showed that the result of the measurement is represented by the quantity

$$Q = \text{Tr}\left(\hat{M}\hat{\Gamma}\right) \tag{10.2}$$

where $\hat{M}$ is a nonnegative-definite Hermitian operator that characterizes the measuring apparatus, and $\hat{\Gamma}$ is the two-point correlation function of the incident light, viewed as an operator. The instrument function itself is a coordinate representation of the measurement operator $\hat{M}$—for example, its matrix element or its Weyl transform. The Weyl transform maps an operator to a Wigner function (for a discussion of the Wigner–Weyl formalism in optics, see Littlejohn and Winston, 1995).

It is appropriate to associate Q with the signal. We then derived an analytical form for the instrument function for a simple radiometer in one space dimension.

One difficulty in using Equation (10.2) is that it may not be easy to compute the instrument function. Although the one-dimensional calculation in Littlejohn and Winston (1995) was not too hard, we do not expect it to be easy to compute the instrument function for many realistic radiometers, which are two-dimensional in cross section and which may have complicated geometry. Therefore, we have considered other means for determining the instrument function. In a previous publication (Winston & Littlejohn, 1997) we considered the possibility that the instrument function could be measured. In this chapter we present an alternative approach. That is, we point out a physical interpretation of the instrument function that is similar to the van Cittert–Zernike theorem. We do this initially by working through the example of a simple "pinhole" radiometer, and then we comment about generalizations.

Radiance is the power per unit volume in phase space. Therefore, an instrument for measuring radiance (called a radiometer) has to select a window function in phase space. For measurements close to the diffraction limit, the exact shape of the window function is not critical. For this reason we examine a simple radiometer, illustrated in Figure 10.2. The dotted line in the figure is the axis of the radiometer. Light enters from the left and passes through the circular pinhole of radius a. It then passes through a drift space of length L, before passing through another circular aperture of radius b. We assume $L \gg a, b$, so the rays are paraxial. The detector is assumed to measure the total power passing through the aperture b and can be thought of as composed of tiny, densely packed, independent absorbing particles (which is a fairly

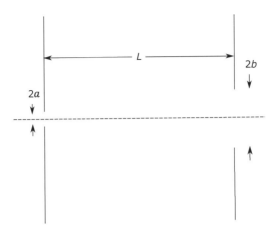

FIGURE 10.2 A pinhole radiometer. The dotted line is the axis. Light enters from the left, passing through circular pinhole a, drift space of length L, and finally circular aperture b.

good approximation to what commonly used detectors like photon detectors, thermal detectors, or photographic film do).

As explained in Littlejohn and Winston (1995) the effect of the radiometer on the radiation field is described by the operator

$$\hat{P} = \hat{A}(b)\hat{D}(L)\hat{A}(a) \tag{10.3}$$

which maps the wave field at the entrance aperture a into the wave field at the exit aperture b. Here $\hat{A}(a)$ is the aperture or "cookie cutter" operator representing the pinhole, $\hat{D}(L)$ is the Huygens-Fresnel operator representing the drift space, and $\hat{A}(b)$ is the aperture operator for the aperture b. In the approximation $L \gg \lambda$ the drift operator has the kernel (or matrix element).

$$\vec{r}_\perp \left| \hat{D}(L) \right| \vec{r}'_\perp = -\frac{ikL}{2\pi}\frac{e^{ikR}}{R^2} \tag{10.4}$$

where $\vec{r}_\perp = (x,y)$, $\vec{r}'_\perp = (x',y')$, $\vec{r} = (x,y,z)$, $\vec{r}' = (x',y',z')$, $L = z - z'$ and $R = |\vec{r} - \vec{r}'|$. Here z is the coordinate along the optical axis and it is assumed that $z > z'$. If in addition we assume rays are paraxial ($r, r' \ll L$), then the kernel can be written

$$\vec{r}_\perp \left| \hat{D}(L) \right| \vec{r}'_\perp = -\frac{ikL}{2\pi} e^{ikL} \exp\left(\frac{ik}{2L} \left| \vec{r}_\perp - \vec{r}'_\perp \right|^2 \right) \tag{10.5}$$

Following the methods of Littlejohn and Winston (1995), the instrument is represented by the operator $\hat{M} = \hat{P}^\dagger \hat{P}$ whose matrix elements are

$$\vec{r}_\perp \left| \hat{M}(L) \right| \vec{r}'_\perp = \left(\frac{kL}{2\pi} \right)^2 \int\limits_{r''_\perp \le b} d^2\vec{r}''_\perp \frac{\exp\left[ik(R_1 - R_2) \right]}{R_1^2 R_2^2} \tag{10.6}$$

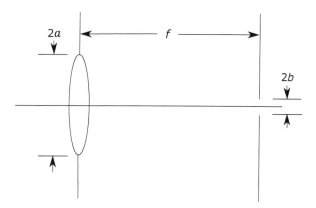

FIGURE 10.3 A practical radiometer. The pinhole is replaced by a lens. The drift space length is essentially the focal length for distant objects $L \approx F$.

where $R_1 = |\vec{r}' - \vec{r}''|$, $R_2 = |\vec{r} - \vec{r}''|$, and the integration is over the detector area. In Equation (10.6) the transverse variables $\vec{r}_\perp$ and $\vec{r}'_\perp$ are understood to lie in the entrance plane (the pinhole), so that $r_\perp, r'_\perp \leq a$. If this condition is not met, the matrix element is understood to be zero.

The expression (10.6) is identical (up to constants) to the mutual intensity evaluated at aperture a of a uniform, delta-correlated source at aperture b (the location of the detector). Thus, to form the van Cittert–Zernike interpretation of the instrument function, we replace the detector (at aperture b, in this example) by a delta-correlated source, and measure the radiation field emanating from the entrance aperture of the instrument (the pinhole in this example). The instrument function (at a given plane) is then proportional to the amplitude at $\vec{r}'$ formed by a spherical wave converging to $\vec{r}$ and diffracted by an aperture the same size, shape, and location as the detector. The detector emulates a delta-correlated source. In a sense, this model involves running the radiometer backwards (exchanging the detector for a source).

This interpretation applies also to other radiometers, for example, those with lenses (see Figure 10.3). The essential property is that the operator $\hat{P}^+$ should serve as a propagator for wave fields traveling to the left (in the negative z direction), just as $\hat{P}$ serves as a propagator for waves travelling to the right. The situation is rather much like time reversal in quantum mechanics. Not all time evolutions in quantum mechanics are time reversal invariant (only those for which the Hamiltonian commutes with time reversal). In the case of optical fields, it is a kind of "z-reversal" that we need. Lenses, drift spaces, and apertures are "z-reversal invariant," as long as evanescent waves can be ignored. We remark that the same conditions apply to the usual van Cittert–Zernike theorem.

10.4   NEAR-FIELD AND FAR-FIELD LIMITS

It is useful to examine the pinhole radiometer and the formalism we have presented in two limiting cases. First, however, we explain some notation regarding the correlation operator and function (see Littlejohn and Winston, 1995, for more details). The mutual coherence is defined by $\Gamma(\vec{r},\vec{r}') = \overline{\psi(\vec{r})\psi^*(\vec{r})}$, where the overbar means a statistical or ensemble average, and where we use a scalar model for the wave field ψ, which in electromagnetic applications can be loosely identified with one of the components of the electric field. When $z=z'$ we can associate the mutual coherence with an operator $\Gamma\Gamma(z)$ by $\Gamma(\vec{r}_\perp,z;\vec{r}'_\perp,z) = \left\langle \vec{r}_\perp \left| \hat{\Gamma}(z) \right| \vec{r}'_\perp \right\rangle$, so that $\hat{\Gamma}(z) = \overline{|\psi(z)\rangle\langle\psi(z)|}$. Thus,

$$\mathrm{Tr}\hat{\Gamma}(z) = \int d^2\vec{r}_\perp \overline{\psi(\vec{r}_\perp,z)^2} \tag{10.7}$$

Thus, if we identify $|\psi|^2$ with 4π times the energy density, then for paraxial rays $(c/4\pi)\mathrm{Tr}\hat{\Gamma}(z)$ is the power crossing the plane z.

Now let us consider the case that a uniform, thermal source is very close to the radiometer, which is useful for normalizing the signal. Then at the entrance aperture the mutual intensity is proportional to a delta function, $\Gamma(\vec{r},\vec{r}') = I_0\lambda^2\delta(\vec{r}_\perp - \vec{r}'_\perp)$, where I_0 is a constant with dimensions of energy/vol. A thermal source is not really delta-correlated, of course; the spatial correlation is really a sine function with a width of the order of a wavelength. It is for this reason that we insert the factor of l2 into the formula for the mutual coherence, so that if we need to set $\vec{r}_\perp = \vec{r}'_\perp$ for the purposes of taking the trace, we can interpret $\lambda^2\delta(0)$ as being of order unity. In any case, when we compute the signal according to Equation (10.2), we obtain

$$\mathrm{Tr}\left(\hat{M}\hat{\Gamma}\right) = I_0\lambda^2\mathrm{Tr}\hat{M} = I_0\lambda^2 N_i \tag{10.8}$$

where we set

$$\mathrm{Tr}\hat{M} = N_i \tag{10.9}$$

for the number of phase space cells in the acceptance region of the instrument.

The trace of $\hat{M}$ is easy to compute. We first make the paraxial approximation in Equation (10.6), which gives

$$\vec{r}_\perp \left| \hat{M} \right| \vec{r}'_\perp = \left(\frac{k}{2\pi L}\right)^2 \int d^2\vec{r}''_\perp \exp\left[\frac{ik}{L}\vec{r}''_\perp \cdot (\vec{r}_\perp - \vec{r}'_\perp)\right] \tag{10.10}$$

where the $\vec{r}''_\perp$ integration is taken over a circle of radius b. Now setting $\vec{r}_\perp = \vec{r}'_\perp$ and integrating $\vec{r}_\perp$ over a circle of radius a, we obtain

$$N_i = \mathrm{Tr}\hat{M} = \left(\frac{kab}{2L}\right)^2 = \left(\frac{\pi a\theta_0}{\lambda}\right)^2 \tag{10.11}$$

where $\theta_0 = b/L$.

The number Ni has a simple interpretation. A phase space cell in the four-dimensional $\vec{k}_\perp - \vec{r}_\perp$ phase space has volume $(2\pi)^2$. As viewed from the standpoint of the exit aperture b of the radiometer, the rays passing through each point of the aperture

b occupy a solid angle of $\pi\left(\dfrac{a}{L}\right)^2$, or a region of $\vec{k}$-space of area $\pi\left(\dfrac{ka}{L}\right)^2$. The region of $\vec{r}$-space is just the exit aperture of area πb^2. Multiplying these areas and dividing by $(2\pi)^2$ gives precisely N_i.

Next we consider the case of a very distant thermal source, which effectively produces a coherent plane wave at the entrance aperture, say, $\sqrt{I_0}\sqrt{\pi\theta_s^2}\,e^{ikz}$, so that $\vec{r}_\perp\left|\hat{\Gamma}(z)\right|\vec{r}_\perp' = I_0\left(\pi\theta_s\right)^2$. The dimensionless factor $\pi\theta_s^2$ will be explained below. Using this and Equation (10.10), we obtain

$$Q = \mathrm{Tr}\left(\hat{M}\hat{\Gamma}\right)$$

$$= I_0\pi\theta_s^2\left(\frac{k}{2\pi L}\right)^2\int d^2\vec{r}_\perp d^2\vec{r}_\perp' d^2\vec{r}_\perp'' \exp\left(\frac{ik}{L}\vec{r}_\perp\cdot\vec{r}_\perp''\right)\exp\left(-\frac{ik}{L}\vec{r}_\perp'\cdot\vec{r}_\perp''\right) \tag{10.12}$$

It is easiest to do the $\vec{r}_\perp$ and $\vec{r}_\perp'$ integrals first (both of which go out to radius a). These are identical, and are given by

$$\int d^2 \exp\left(\pm\frac{ik}{L}\vec{s}\cdot\vec{r}_\perp''\right) = \pi a^2 F\left(\frac{ka\vec{r}_\perp''}{L}\right) \tag{10.13}$$

where $s = \vec{r}_\perp$ or $\vec{r}_\perp'$. This leaves only the $\vec{r}_\perp''$ integration (taken out to radius b).

$$Q = \mathrm{Tr}\left(\hat{M}\hat{\Gamma}\right) = I_0\pi\theta_s^2\left(\frac{\pi a^2}{\lambda L}\right)^2\int d^2\vec{r}_\perp''\left[\frac{2J_1\left(\dfrac{ka\vec{r}_\perp''}{L}\right)}{\dfrac{ka\vec{r}_\perp''}{L}}\right] \tag{10.14}$$

which agrees with the expected fraction of the Airy diffraction pattern contained by the detector of radius b.

10.5 A WAVE DESCRIPTION OF MEASUREMENT

We begin with the statistical properties of the incident wave field. The two-point correlation function at the source plane is

$$\vec{r}_\perp\left|\hat{\Gamma}(0)\right|\vec{r}_\perp' = I_0\delta\left(\vec{r}_\perp - \vec{r}_\perp'\right) \tag{10.15}$$

with $\vec{r}_\perp$ and $\vec{r}_\perp'$ inside the source, σ, and equal to 0 otherwise. The matrix elements of the Fresnel free space propagator of a distance L are given by

$$\vec{r}_\perp\left|\hat{D}(L)\right|\vec{r}_\perp' = -\frac{ik}{2\pi L}e^{ikL}_{0}\exp\left(\frac{ik}{2L}\left|\vec{r}_\perp - \vec{r}_\perp'\right|^2\right) \tag{10.16}$$

where $k = 2\pi/\lambda$. The two-point correlation at a plane a distance L from the source in the Fresnel diffraction regime is

$$\left\langle \vec{r}_\perp \left| \hat{\Gamma}(L) \right| \vec{r}_\perp' \right\rangle = I_0 \int d^2\vec{s}_\perp d^2\vec{s}_\perp' \left\langle \vec{r}_\perp \left| \hat{D}(L) \right| \vec{s}_\perp \right\rangle \vec{s}_\perp \left\langle \hat{\Gamma}(0) \right| \vec{s}_\perp' \right\rangle \vec{s}_\perp' \left\langle \hat{D}(L) \right| \vec{r}_\perp' \right\rangle$$

$$= I_0 \int d^2\vec{s}_\perp \left\langle \vec{r}_\perp \left| \hat{D}(L) \right| \vec{s}_\perp \right\rangle \vec{s}_\perp \left\langle \vec{s}_\perp \left| \hat{D}(L) \right| \vec{r}_\perp' \right\rangle$$

$$= I_0 \exp\left(\frac{ik}{2L} |\vec{r}_\perp - \vec{r}_\perp'|^2 \right) \left(\frac{ik}{2\pi L} \right)^2 \tag{10.17}$$

$$\times \int_\sigma d^2\vec{s}_\perp \exp\left(\frac{ik}{L} \vec{s}_\perp \cdot (\vec{r}_\perp - \vec{r}_\perp') \right)$$

$$= I_0 \exp\left(\frac{ik}{2L} (|\vec{r}_\perp|^2 - |\vec{r}_\perp'|^2) \right) F_r(\vec{r}_\perp - \vec{r}_\perp')$$

In Equation (10.17), $F_r(\vec{r}_\perp - \vec{r}_\perp')$ is the Fourier transform of the source area; it is the VCZ result for the far-field two-point correlation function.

We will be testing two source geometries, circular and square. For the square geometry F_r is

$$F_r(\vec{r}_\perp - \vec{r}_\perp') = \left(\frac{1}{\pi} \right)^2 \frac{1}{(x-x')(y-y')} \sin\left(\frac{kd}{2L}(x-x') \right) \sin\left(\frac{kd}{2L}(y-y') \right) \tag{10.18}$$

with d the linear dimension of the source. For the circular geometry F_r is

$$F_r(\vec{r}_\perp - \vec{r}_\perp') = \frac{kd}{2L} \frac{J_1\left(\frac{kd}{2L} |\vec{r}_\perp - \vec{r}_\perp'| \right)}{2\pi |\vec{r}_\perp - \vec{r}_\perp'|} \tag{10.19}$$

10.6 FOCUSING AND THE INSTRUMENT OPERATOR

The matrix elements for the lens operator are given by

$$\left\langle \vec{r}_\perp \left| \hat{L}(f) \right| \vec{r}_\perp' \right\rangle = \exp\left(-\frac{ik}{2f} |\vec{r}_\perp|^2 \right) \delta(\vec{r}_\perp - \vec{r}_\perp') \tag{10.20}$$

The instrument operator is

$$\hat{M} = \hat{P}^\dagger \hat{P} \tag{10.21}$$

where $\hat{P}$ is the propagator from the aperture, $\hat{A}(a)$, to the detector $\hat{A}(b)$.

$$\hat{P} = \hat{A}(b)\hat{D}(l)\hat{L}(f)\hat{A}(a) \tag{10.22}$$

The matrix elements for $\hat{A}(a)$ are given by

$$\left\langle \vec{r}_\perp \left| \hat{A}(a) \right| \vec{r}_\perp' \right\rangle = \delta(\vec{r}_\perp - \vec{r}_\perp') \tag{10.23}$$

with $\vec{r}_\perp$ and $\vec{r}'_\perp$ inside the aperture, a, and equal to 0 otherwise. The matrix elements for $\hat{A}(b)$ are given by

$$\vec{r}_\perp \left| \hat{A}(b) \right| \vec{r}'_\perp = \delta\left(\vec{r}_\perp - \vec{r}''_\perp\right) \tag{10.24}$$

with $\vec{r}_\perp$ and $\vec{r}'_\perp$ inside the detector, b, and equal to 0 otherwise.

From Equations (10.16) and (10.20) we have

$$\vec{r}_\perp \left| \hat{D}(l)\hat{L}(f) \right| \vec{r}'_\perp = -\frac{ik}{2\pi l} e^{ikl} \exp\left(\frac{ik}{2l}\left|\vec{r}_\perp - \vec{r}'_\perp\right|^2\right) \exp\left(-\frac{ik}{2f}\left|\vec{r}'_\perp\right|^2\right) \tag{10.25}$$

This gives

$$\vec{r}_\perp \left| \hat{P} \right| \vec{r}'_\perp = -\frac{ik}{2\pi l} \exp\left(\frac{ik}{2l}\left|\vec{r}_\perp - \vec{r}'_\perp\right|^2\right) \exp\left(-\frac{ik}{2f}\left|\vec{r}'_\perp\right|^2\right) \tag{10.26}$$

with $\vec{r}_\perp$ inside the detector, $\vec{r}'_\perp$ inside the aperture, and equal to 0 otherwise.

The matrix elements of $\hat{M}$ are given by

$$\vec{r}_\perp \left| \hat{M} \right| \vec{r}'_\perp = \int_b d^2\vec{s}_\perp \vec{s}_\perp \left| \hat{P} \right| \vec{r}_\perp * \vec{s}_\perp \left| \hat{P} \right| \vec{r}'_\perp \tag{10.27}$$

The integration is over the detector area.

$$\vec{r}_\perp \left| \hat{M} \right| \vec{r}'_\perp = \exp\left(\frac{ik}{2f}\left(\left|\vec{r}_\perp\right|^2 - \left|\vec{r}'_\perp\right|^2\right)\right) \exp\left(-\frac{ik}{2l}\left(\left|\vec{r}_\perp\right|^2 - \left|\vec{r}'_\perp\right|^2\right)\right) \left(\frac{k}{2\pi l}\right)^2$$

$$\times \int_b d^2\vec{s}_\perp \exp\left(\frac{ik}{l}\vec{r}_\perp \cdot (\vec{r} - \vec{r}')\right) \tag{10.28}$$

$$= \exp\left(\frac{ik}{2f}\left(\left|\vec{r}_\perp\right|^2 - \left|\vec{r}'_\perp\right|^2\right)\right) \exp\left(\frac{-ik}{2l}\left(\left|\vec{r}_\perp\right|^2 - \left|\vec{r}'_\perp\right|^2\right)\right) F_i\left(\vec{r}_\perp - \vec{r}'_\perp\right) \tag{10.29}$$

with $\vec{r}_\perp$ and $\vec{r}'_\perp$ inside the aperture, a, and equal to 0 otherwise. Here F_i is

$$F_i\left(\vec{r}_\perp - \vec{r}'_\perp\right) = \left(\frac{k}{2\pi l}\right)^2 \int_b d^2\vec{s}_\perp \exp\frac{ik}{l}\vec{s}_\perp \cdot (\vec{r} - \vec{r}') \tag{10.30}$$

If the camera is focused on the source

$$\frac{1}{l} + \frac{1}{L} = \frac{1}{f} \tag{10.31}$$

where L is the distance between the camera entrance aperture and the source. In this case the nonzero matrix elements are given by

$$\vec{r}_\perp \left| \hat{M} \right| \vec{r}'_\perp = \exp\left(\frac{ik}{2l}\left(\left|\vec{r}_\perp\right|^2 - \left|\vec{r}'_\perp\right|^2\right)\right) F_i\left(\vec{r}_\perp - \vec{r}'_\perp\right) \tag{10.32}$$

The VCZ result for the nonzero matrix elements of $\hat{M}$ are given by Equation (10.30) with the camera focused at infinity, $l = f$. This results in a slight difference between the VCZ result and the general result in time phase space volume of $\hat{M}$ defined by

$$N_i = \text{Tr}\left(\hat{M}\right) \tag{10.33}$$

with N_i being time number of phase space cells. The infrared camera we used in the experiment has a circular entrance aperture and a square detector. We thus have

$$F_i\left(\vec{r}_\perp - \vec{r}_\perp'\right) = \left(\frac{1}{\pi}\right)^2 \frac{1}{\left(x - x'\right)\left(y - y'\right)} \sin\left(\frac{kb}{l}\left(x - x'\right)\right)\sin\left(\frac{kb}{l}\left(y - y'\right)\right) \tag{10.34}$$

with $\vec{r}_\perp$ and $\vec{r}_\perp'$ inside the circular entrance aperture of radius a and equal to zero otherwise. The linear dimension of the square detector is $2b$. From Equation (10.32),

$$N_i = \pi a^2 \left(\frac{2\theta_i}{\lambda}\right)^2 \tag{10.35}$$

where $2\theta_i = 2b/l$ is the full acceptance angle. For all practical purposes, the difference in the values of N_i for the focused case and with $l = f$ is negligible. Henceforth, we will mean $l = f$ when referring to N_i.

10.7   MEASUREMENT BY FOCUSING THE CAMERA ON THE SOURCE

From Equations (10.17) and (10.32), the detected signal, Q, is

$$Q = \text{Tr}\left(\hat{M}\hat{\Gamma}\left(L\right)\right) = I_0 \int d^2\vec{r}_\perp d^2\vec{r}_\perp' F_i\left(\vec{r}_\perp - \vec{r}_\perp'\right) F_r\left(\vec{r}_\perp - \vec{r}_\perp'\right) \tag{10.36}$$

The normalized signal, Q_n, is defined as

$$Q_n = \frac{Q}{I_0 N_i} \tag{10.37}$$

10.8   EXPERIMENTAL TEST OF FOCUSING

The matrix elements of the instrument operator modeling the infrared camera used in the experiment are given by Equation (10.32). The experiment was conducted by focusing the camera on the source. The results of the experiments are compared with Equations (10.36) and (10.37).

The following is the protocol for processing the data. The value of the signal when the detector of the camera is flood-illuminated by the blackbody radiation subtracted by the value of the signal when the detector is flood-illuminated by the background is used as the normalization. The measured normalized signal is obtained by first subtracting the detected signal from the background and then dividing by the normalization. The normalized signal is compared with theory.

The camera has an interference filter and a HgCdTd detector giving a wavelength window with a peak at $\lambda=8.8$ mm and $\Delta\lambda=\pm 0.75$ mm. The square HgCdTd detector is of dimensions 75 mm $\times$ 75 mm. The focal length of the camera is $f=18.99$ mm. In Equation (10.32) Fi is then given by Equation (10.30), with $b=(75/2)$ μm.

Two circular aperture plates of radii $a=0.136$in and $a=0.272$in were placed in front of the camera aperture. We have $N_i=1.888$ for the a=0.136 in aperture and $N_i=7.551$ for the $a=0.272$ in aperture. The face of each plate facing the lens was painted with high emissive paint to provide the background. Time, source, size, and shape were controlled by placing aluminum masks with either square or circular apertures over a blackbody source set at T ~ 500°C.

Define Nr as

$$N_r = \int_a d^2 \vec{r}_\perp \vec{r}_\perp \left| \hat{\Gamma}\left(L\right) \right| \vec{r}_\perp' \tag{10.38}$$

where the integration is over the camera aperture. Nr can be interpreted as the number of phase space cells of the radiation field intercepted by the camera aperture. Therefore, Nr can be much less than one without violating the uncertainty principle. For a square source

$$N_r = \pi \left(\frac{2a\theta_s}{\lambda} \right)^2 \tag{10.39}$$

and for a circular source

$$N_r = \left(\frac{\pi a\theta_s}{\lambda} \right)^2 \tag{10.40}$$

Here θ_s is the half angle subtended by the source at the camera's entrance aperture. For the square source $\theta_s=d/2L$ with d being the linear dimension for the square source. For the circular source $\theta_s=d/2L$ with d being the diameter of the circular source.

The results of the measured normalized signal are plotted with theory versus N_r in Figures 10.4 and 10.5. The agreement with the experiment is highly satisfactory.

10.9   CONCLUSION

We have demonstrated a consistent approach to incorporating the diffraction properties of the instrument in optical measurements. We used a remarkable analogy between the result of measuring radiance and the van Cittert–Zernike theorem that exploits the symmetry between an incoherent source whose radiance is being measured and the detector whose signal represents the measurement. It turns out that the measured radiance is represented (up to an overall constant) by the double integral over time instrument aperture of the mutual intensity of the field and the mutual

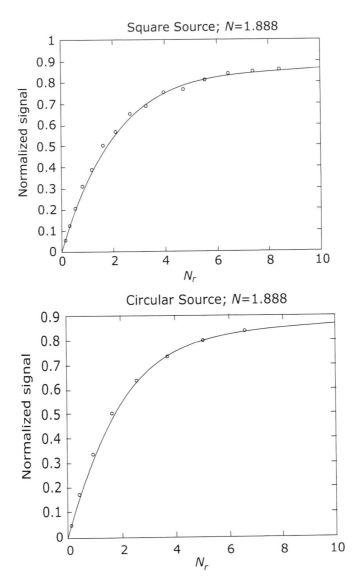

FIGURE 10.4 Comparison of experiment with theory for the $N = 1.888$ peak measurements. Filled circles are data points.

intensity of a delta-correlated source the same size, shape, and location as the detector. While we have expressed our results in a time context of radiometry, one would go through a similar analysis in analyzing the detection of any partially coherent wave. The signal is represented by the double integral of two mutual coherence functions. One of these is for the incident wave, the other arising from the detector considered as a source. It is likely that entirely similar considerations may apply to other signal detection processes where diffraction effects are important.

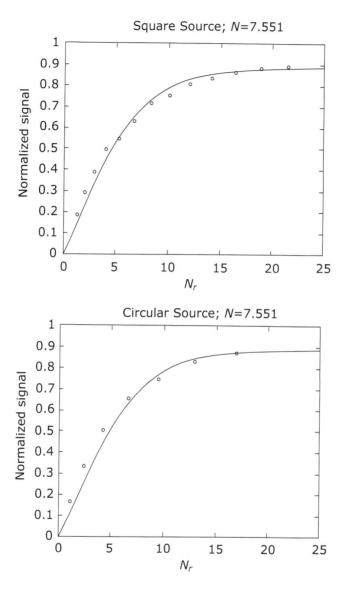

FIGURE 10.5 Comparison of experiment with theory for the $N=7.551$ peak measurements. Filled circles are data points.

REFERENCES

Born, M., and Wolf, E. (1999). *Principles of Optics*, 7th Ed. Cambridge University Press.

Littlejohn, R. G., and Winston, R. (1993). Corrections to classical radiometry. *J. Opt. Soc. Am. A* **10**, 2024–2037.

Littlejohn, R. G., and Winston, R. (1995). Generalized radiance and measurement. *J. Opt. Soc. Am. A* **12**, 2736–2743.

Mandel, L., and Wolf, E. (1995). *Optical Coherence and Quantum Optics*. Cambridge University Press.

Mather, J. C., Toral, M., and Hemmati, H. (1986). Heat trap with flare as multi-mode antenna. *Appl. Opt.* **25**, 2826–2830.

Sun, Y., Winston, R., O'Gallagher, J. J., and Snail, K. A. (2002). Statistical optics and radiance measurement in the diffraction limit. *Opt. Commun.* **206**, 243–251.

Walther, A. (1973). Radiometry and coherence. *J. Opt. Soc. Am. A* **63**, 1622–1623.

Winston, R., and Littlejohn, R. G. (1997). Measuring the instrument function of radiometers. *J. Opt. Soc. Am. A* **14**, 3099–3101.

Winston, R., Sun, Y., and Littlejohn, R. G. (2002). Measuring radiance and the van Cittert–Zernike theorem. *Opt. Commun.* **207**, 41–48.

Wolf, E. (1978). Coherence and radiometry. *J. Opt. Soc. Am. A* **68**, 6–17.

Appendix A
Derivation and Explanation of the Étendue Invariant, Including the Dynamical Analogy; Derivation of the Skew Invariant

A.1 THE GENERALIZED ÉTENDUE

In Section 2.7 we introduced the invariant

$$n^2 dx \, dy \, dL \, dM \tag{A.1}$$

The meaning of this was as follows: let any ray be traced through an optical system, and let rectangular coordinate axes be set up in the entry and exit spaces in arbitrary orientations. Also, let the ray meet the x, y plane in the entry space in (x, y), and let its direction cosines be (L, M, N), and similarly for the exit space. Then for any nearby ray with coordinates $(x + dx, y + dy, L + dL, M + dM)$ we have

$$n'^2 dx' dy' dL' dM' = n^2 dx \, dy \, dL \, dM \tag{A.2}$$

We shall prove in Section A.2 that this result is true for any optical system and for any choice of the directions of the axes. But first we discuss some different ways of interpreting this result. We put $p = nL$ and $q = nM$, and we treat (x, y, p, q) as coordinates in a four-dimensional space. Let U be any enclosed volume in this space. Then U is given simply by

$$\int dU = \int\int\int\int dx \, dy \, dp \, dq \tag{A.3}$$

taken over the volume. The result of Equation (A.2) means that if we let a ray sweep out the boundary of this volume in the four-dimensional entry space, it will sweep out an equal volume in four-dimensional exit space or, clearly, in any intermediate space of the optical system. There is a useful analogy between this space and the multidimensional phase space of Hamiltonian mechanics, so we shall call it simply phase space. Thus, our result is that if a ray sweeps out a certain volume of phase space in the entry region, it sweeps out the same total volume in the exit region, or, more concisely, phase space volume is conserved. Of course, the shape of the volume will change from entry to exit space or even if the origin of coordinates is shifted parallel to the z direction. However, it can easily be shown that rotating the axes about their origin does not change the phase space volume, so it is a physical significant invariant.

Another picture that is particularly valuable for the purposes of this book may be obtained by supposing that an area in the x, y plane (in real space, not phase space) is covered with uniformly closely spaced points and that through each point a large number of rays is drawn. These rays are to be in different directions so that their direction cosines increase by uniformly small increments over certain ranges of L and M. The rays so drawn then represent "ray points" uniformly spaced throughout a certain volume of phase space, and our theorem can then be stated in the following form: the density of rays in phase space is invariant through the optical system. If the four-volume U of Equation (A.3) has been defined by means of physical components in the optical system—that is, an aperture stop and a field stop—it is known as the throughput or étendue of the system.

We can see from the ray-point representation that the étendue is a measure of the power that can be transmitted through the optical system from a uniformly bright source of sufficient extent, in the geometrical optics approximation. Here, of course, in addition to neglecting interference and diffraction effects, we are assuming no losses in the system due to reflections, absorption, or scattering. The concept of étendue is applied in comparing the properties of many different kinds of optical instruments, although usually only a simplified version appropriate to axially symmetric systems is used.

A.2 PROOF OF THE GENERALIZED ÉTENDUE THEOREM

Equation (A.2) has been proved in many different ways. In particular, the analogy from Hamiltonian mechanics can be used to give a simple proof (see, e.g., Marcuse, 1972; Winston, 1970). However, since we are here concerned with applications to geometrical optics, it seems appropriate to give a proof based directly on optical principles. We shall discuss the mechanical analogies in some detail in Section A.3.

We follow closely the method of Welford (1974), which makes use of the pointeikonal, or characteristic function, of Hamilton. This function is defined as follows. We take arbitrary cartesian coordinate systems in entry and exit spaces of the optical system under discussion. Let P and P' be any two points in the entry and exit spaces, respectively, and let them be in the x, y planes of their respective coordinate system.[*] Then the eikonal V is defined as the optical path length from P to P' along the physically possible ray joining them. In general, one and only one ray passes through P and P', but if there is more than one, then V is multivalued. Thus, with the preceding restriction V is a function of x, y, x', and y'. Let the direction cosines of the ray in the two spaces be (L, M, N) and (L', M', N'); then the fundamental property of the eikonal can be stated as follows:

$$\partial v / \partial x = -nL, \ \ \partial v / \partial x = -nM$$

$$\partial v / \partial x' = -n'L', \ \ \partial v / \partial x' = -n'M'$$

(A.4)

This property is proved in many texts on geometrical optics (e.g., Born & Wolf, 1975; Welford, 1974). To prove our theorem we differentiate Equation (A.4) again, and we obtain, using the notations of Equation (A.3) and using subscripts for partial derivatives,

$$dp = -V_{xx}dx - V_{xx}dy - V_{xx'}dx' - V_{xy'}dy'$$

$$dq = -V_{yx}dx - V_{yx}dy - V_{yx'}dx' - V_{yy'}dy'$$

$$dp' = -V_{x'x}dx + V_{x'x}dy + V_{x'x'}dx' + V_{x'y'}dy'$$

$$dq' = -V_{y'x}dx + V_{y'x}dy + V_{y'x'}dx' + V_{y'y'}dy'$$

(A.5)

We next rearrange these terms and put the equations in matrix form

$$\begin{pmatrix} V_{xx'} & V_{xy'} & 0 & 0 \\ V_{yx'} & V_{yy'} & 0 & 0 \\ V_{x'x'} & V_{x'y'} & -1 & 0 \\ V_{y'x'} & V_{y'y'} & 0 & -1 \end{pmatrix} \begin{pmatrix} dx' \\ dy' \\ dp' \\ dq' \end{pmatrix} = \begin{pmatrix} -V_{xx} & -V_{xy} & -1 & 0 \\ -V_{yx} & -V_{yy} & 0 & -1 \\ -V_{x'x} & -V_{x'y} & 0 & 0 \\ -V_{y'x} & -V_{y'y} & 0 & 0 \end{pmatrix} \begin{pmatrix} dx \\ dy \\ dp \\ dq \end{pmatrix}$$

(A.6)

If we denote the two matrices by B and A and the column vectors by M and M', this equation takes the form

$$BM' = AM$$

(A.7)

and multiplying through by the inverse of B,

$$M' = B^{-1}AM$$

(A.8)

This matrix equation can be expanded, and we get together with three similar equations

$$dx' = \frac{\partial x'}{\partial x}dx + \frac{\partial x'}{\partial y}dy + \frac{\partial x'}{\partial p}dp + \frac{\partial x'}{\partial q}dq$$

(A.9)

It can be seen that the determinant of the matrix $B^{-1}A$ is the Jacobian

$$\det\left(B^{-1}A\right) = \frac{\partial\left(x', y', p', q'\right)}{\partial\left(x, y, p, q\right)}$$

(A.10)

which transforms the differential four-volume dx, dy, dp, dq—that is, we have

$$dx'dy'dp'dq' = \frac{\partial\left(x', y', p', q'\right)}{\partial\left(x, y, p, q\right)} dx\, dy\, dp\, dq$$

(A.11)

Our result will be proven if we can show that the Jacobian has the value unity. But the determinant of matrix B has the value

$$V_{xx'}V_{yy'} - V_{xy'}V_{yx'}$$

(A.12)

and that of matrix A has the same value, since $V_{xy'} = V_{y'x}$, and so forth. Also, the determinant of the product of two square matrices is the product of their determinants, so that

$$\det\left(B^{-1}\right) = \left(V_{xx'}V_{yy'} - V_{xy'}V_{yx'}\right)^{-1} \tag{A.13}$$

Thus, $\det\left(B^{-1}A\right) = 1$ and Equation (A.13) yields our theorem, Equation (A.2).

A.3   THE MECHANICAL ANALOGIES AND LIOUVILLE'S THEOREM

In this section we shall indicate the analogies used to identify our theorem of the invariance of U, the étendue, with Liouville's theorem in statistical mechanics.

Fermat's principle, on which all of geometrical optics can be based, can be stated in the form

$$\delta \int_{P_1}^{P_2} n\left(x, y, z\right) ds = 0 \tag{A.14}$$

where ds is an element of the ray path from P_1 to P_2. This can be written in the form

$$\delta \int_{P_1}^{P_2} \mathcal{L}\left(x, y, \dot{x}, \dot{y}\right) = 0 \tag{A.15}$$

where

$$\mathcal{L}\left(x, y, \dot{x}, \dot{y}\right) = n\left(x, y, z\right)\sqrt{1 + \dot{x}^2 + \dot{y}^2} \tag{A.16}$$

and the dots denote differentiation with respect to z. Also, we define

$$p = \frac{n\dot{x}}{\sqrt{1 + \dot{x}^2 + \dot{y}^2}} \quad q = \frac{n\dot{y}}{\sqrt{1 + \dot{x}^2 + \dot{y}^2}} \tag{A.17}$$

The analogy is to regard L as the Lagrangian function of a mechanical system in which x and y are two generalized coordinates, p and q are the corresponding generalized momenta, and z corresponds to the time axis. On this basis the ordinary development of mechanics can be carried out, such as by Luneburg (1964, Article 18), by solving the variational problem of Equation (A.15). The Hamiltonian is found to have the value

$$\mathcal{H} = -\sqrt{n^2 - p^2 - q^2} \tag{A.18}$$

In other words, it is $-nN$ where N is the z-direction cosine, and, of course, p and q as just defined are respectively equal to nL and nM. The phase space for this system has the four coordinates (x, y, p, q), and Liouville's theorem in statistical mechanics can be invoked immediately to state that phase space volume is conserved.

However, the meaning of Liouville's theorem in mechanics is rather different from the theorem of conservation of étendue. Liouville's theorem is essentially statistical in nature, and it refers to the evolution in time of an ensemble of mechanical systems of identical properties but with different initial conditions. Each system is represented by a single point in phase space, and the theorem states that the average density of points in phase space is constant in time. An example would be the molecules of a perfect classical gas in equilibrium in a container. Each point in phase space, which in this example has $2N$ dimensions, where N is the number of molecules, represents one of an ensemble of identical containers, an ensemble large enough to permit taking a statistical average of the density of representative points. Liouville's theorem states that if all the containers remain in equilibrium, the average density of points remains constant.

Another example would be focused beams of charged particles, as in a particle accelerator. Here we can regard one pulse of particles as constituting one realization of the ensemble and therefore one point in phase space. The statistical averaging is carried out over the random positions and momenta of the particles entering the focusing system from pulse to pulse.

The theorem has been applied to many different physical systems but the essential point is that it makes a statistical statement about the average density of points in phase space, whereas the throughput or étendue theorem is deterministic in nature. Thus, although for convenience we may call the throughput theorem "Liouville's theorem," it is desirable to remember that there is a fundamental difference in meaning.

A.4 CONVENTIONAL PHOTOMETRY AND THE ÉTENDUE

In discussing the action of concentrators we use the notion of a beam of radiation in which a certain cross section $dx\,dy$ is uniformly filled with rays covering uniformly the direction cosine solid angle $dL\,dM$. Note the useful relation $dL\,dM = N\,d\Omega$. Then an ideal concentrator takes all the rays from a source of finite area within a certain range of direction cosines and delivers them at the exit aperture to emerge over a solid angle 2π.

These ideas relate easily to particle beams, but it may not be quite clear how they relate to classical photometric ideas. In classical photometry we have a particular kind of ideal source called Lambertian for which the flux or power radiated per unit solid angle per unit projected area of the source is constant over all directions. Many physical sources approximate closely to the Lambertian condition and an ideal blackbody radiator must be Lambertian, since the radiation in a blackbody cavity is isotropic. It is easy to show that if a source radiates the same flux per unit area of the source and per unit element of direction cosine space, then it is Lambertian. Thus, we see than an ideal concentrator will take radiation from a Lambertian source over a certain solid angle and deliver it so as to make the exit aperture appear to be a complete Lambertian radiator over solid angle 2π.

REFERENCES

Born, M., and Wolf, E. (1975). *Principles of Optics*, 5th Ed. Pergamon, Oxford.

Marcuse, D. (1972). *Light Transmission Optics*. Van Nostrand-Reinhold, Princeton, NJ.

Welford, W. T. (1974). *Aberrations of the Symmetrical Optical System*. Academic Press, New York.

Winston, R. (1970). Light collection within the framework of geometrical optics. *J. Opt. Soc. Am.* **60**, 245–247.

Appendix B
The Luneburg Lens

The Luneburg lens, discussed in Chapter 3, is not, of course, of any use as a practical concentrator. It is, however, the simplest example of an ideal concentrator with maximum theoretical concentration for collecting angles up to $\pi/2$, and therefore we develop its theory and general properties in this appendix. The main reference is to Luneburg's own work (Luneburg, 1964). Some of the general geometrical optics background can also be found in Born and Wolf (1975), and extensions are given by Morgan (1958) and Cornbleet (1976).

Our starting point is the differential equation of the light rays. Let s be the distance along a ray measured from some fixed origin, let r be the position vector of a point on a ray, and let $n(r)$ be the refractive index, varying continuously as a function of r. Then it can easily be shown (e.g., Born & Wolf, 1975, Section 3.2.1) that the rays are given by solutions of

$$\frac{d}{ds}\left(n(\mathbf{r})\frac{dr}{ds} \right) = \operatorname{grad} n(\mathbf{r}) \qquad (B.1)$$

Let the refractive index distribution have spherical symmetry. Then we can write $n(r) \equiv n(r)$, where r is a radial coordinate from the origin. Also, on account of the spherical symmetry, any light ray lies wholly in a plane through the origin, and we can describe its path by the polar coordinates (r, θ).

In order to do this we note that dr/ds is a unit vector along the tangent to the ray—say, s. Now we have

$$\frac{d}{ds}\left(\mathbf{r} \times ns \right) = \mathbf{s} \times ns + r \times \frac{d(ns)}{ds} = \mathbf{r} \times \operatorname{grad} n$$

from Equation (B.1). But by spherical symmetry grad n is parallel to r and so we have

$$\frac{d}{ds}\left(r \times ns \right) = 0$$

and a first integral of Equation (B.1) is

$$r \times ns = \text{const.}$$

This is a vectorial equation that is equivalent to establishing the conservation of the skew relative to two orthogonal axes.

Since the rays are plane curves as just noted, this gives, say,

$$nr\sin\phi = \text{const.} = h \qquad (B.2)$$

where ϕ is as in Figure B.1. Since

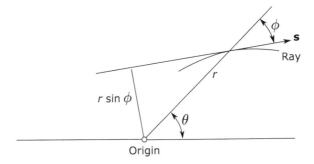

FIGURE B.1 Notation for ray paths in a medium with spherical symmetry.

$$\tan \phi = r \frac{d\theta}{dr}$$

Equation (C.2) yields

$$\frac{d\theta}{dr} = \frac{h}{r\sqrt{n^2 r^2 - h^2}} \tag{B.3}$$

and thus if $n(r)$ is specified, the ray paths are obtained by quadrature

$$\theta_1 - \theta_0 = h \int_{r_0}^{r_1} \frac{dr}{r\sqrt{n^2 r^2 - h^2}} \tag{B.4}$$

If we wish to find an index distribution $n(r)$ that will act as an aberration-free medium, we have to regard Equation (B.4) as an integral equation for the function $n(r)$. Luneburg did this and considered the use of an index distribution inside the unit sphere and with index unity at the surface of the sphere. He formulated the conditions for focusing from any point r_0 to another r_1 as in Figure B.2, and he obtained an explicit solution for the case of present interest, $r_0 = \infty$, $r_1 = 1$.

Luneburg's derivation is very complex and closely argued. Rather than reproduce his work we shall simply verify the solution he gave.

Let the lens have radius a and let the index vary in such a way that r has only one minimum value for the ray path in the lens—say, r^*—as in Figure B.3. Then

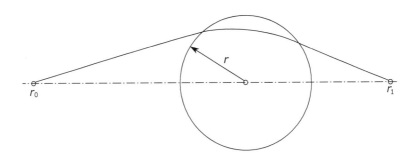

FIGURE B.2 Luneburg's general problem of an aberration-less unit sphere.

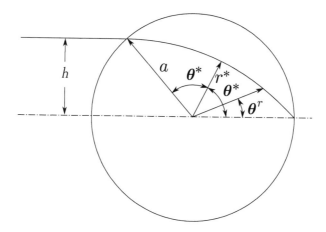

FIGURE B.3   Luneburg's problem for $r_0 = \infty$, $r_1 = 1$.

from Equation (C.2) we have, using values at the beginning of the trajectory in the lens,

$$nr \sin \phi \equiv a\frac{h}{a} = h$$

and thus

$$n\left(r^*\right)r^* = h \tag{B.5}$$

Since the two halves of the trajectory on either side of the radius r^* must be mirror images we have

$$\theta^* = \frac{\pi}{2} - \frac{1}{2}\arcsin\left(\frac{h}{a}\right) \tag{B.6}$$

and thus Equation (C.4) takes the form

$$\frac{\pi}{2} - \frac{1}{2}\arcsin\frac{h}{a} = h\int_{r^*}^{a}\frac{dr}{r\sqrt{n^2 r^2 - h^2}} \tag{B.7}$$

to be satisfied by a function $n(r)$ for any $h < a$.

Luneburg gave the distribution

$$n(r) = \sqrt{2 - \frac{r^2}{a^2}} \tag{B.8}$$

as the solution of this integral equation. If we substitute for $n(r)$, we have to evaluate

$$h = \int_{r^*}^{a}\frac{dr}{r\sqrt{2r^2 - \dfrac{r^4}{a^2} - h^2}} \tag{B.9}$$

to verify the solution. From Equation (C.5) we find

$$r^{*2} = a^2 \left[1 - \sqrt{1 - \frac{h^2}{a^2}} \right] \tag{B.10}$$

and if we make the change of variable

$$r^2 = a^2 \left[1 - \sin\beta \sqrt{1 - \frac{h^2}{a^2}} \right] \tag{B.11}$$

in the integral, it becomes

$$h \int_0^{\pi} \frac{d\beta}{2a \left(1 - \left[1 - \sin\beta \sqrt{1 - \frac{h^2}{a^2}} \right] \right)} \tag{B.12}$$

The indefinite integral has the value

$$\tan^{-1} \left\{ \frac{\tan\dfrac{\beta}{2} - \sqrt{1 - \dfrac{h^2}{a^2}}}{\dfrac{h}{a}} \right\} \tag{B.13}$$

and on substituting the limits of integration and recalling that $h/a = \sin 2\theta^*$, we find that the value is θ^*, as required by Equation (C.7). We have also obtained the actual ray paths, since these are specified by θ as a function of r for given h by

$$\theta_2 - \theta_1 = \tan^{-1} \left\{ \frac{\tan\dfrac{\beta}{2} - \sqrt{1 - \dfrac{h^2}{a^2}}}{\dfrac{h}{a}} \right\} \Bigg|_{r_0}^{r_1} \tag{B.14}$$

With $\sin\beta = \left[1 - \left(\dfrac{r^2}{a^2} \right) \right] \left(1 - \dfrac{h^2}{a^2} \right)^{-\frac{1}{2}}$. Figure B.4 shows several rays plotted to scale

as in Figure B.5. A ray incident at height h from the center emerges at an angle to the surface arcsin(h/a) so that at any point on the exit surface rays emerge in all directions up to $\pi/2$ from the normal and the sines of the angles are distributed uniformly. Thus, the exit surface is filled with rays at all possible angles, and this must be in some sense a system with maximum theoretical concentration ratio. We can make this more explicit by calculating the entering étendue. This is, from Figure B.5, the area $\pi a^2 \sec\theta$ integrated over the direction cosines of the incoming beam—that is, it is

$$\int_0^{\theta_1} \int_0^{2\pi} \pi a^2 \sec\theta \, d\left(\sin\theta \cos\phi \right) d\left(\sin\theta \sin\phi \right)$$

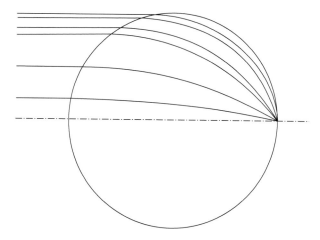

FIGURE B.4   Rays in the Luneburg lens.

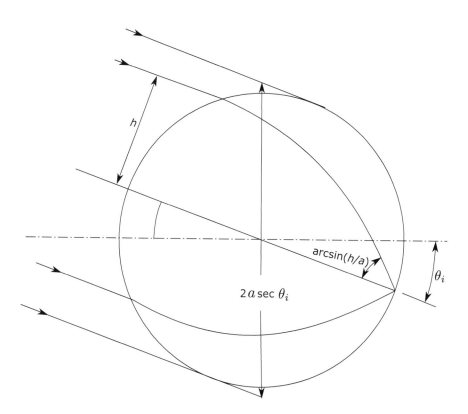

FIGURE B.5   The Luneburg lens as a concentrator of maximum theoretical concentration ratio for any entry angle θ_i less than $\pi/4$.

where ϕ is an azimuthal angle. The value of this is $2\pi^2 a^2 (1 - \cos\theta_i)$. Also, the integral over the direction cosines alone is

$$\int_0^{\theta_1}\int_0^{2\pi} d(\sin\theta\cos\phi)\, d(\sin\theta\sin\phi)$$

and this is $\pi\sin^2\theta_i$. Thus the effective entry area must be $\pi a^2 / \cos^2(1/2)\theta_i$.

The area of the spherical cap that forms the exit surface is $4\pi a^2 \sin^2(1/2)\theta_i$ so that the concentration ratio is $1/\sin^2\theta_i$, the theoretical value for a collecting angle θ_i. Since all entering rays emerge, we have a 3D concentrator with the maximum theoretical concentration ratio.

We had to proceed in this slightly oblique way because the collecting aperture of the lens shifts with angle and because the exit surface is not planar. Thus, the correspondence with systems with well-defined and plane entry and exit apertures is not perfect.

REFERENCES

Born, M., and Wolf, E. (1975). *Principles of Optics*, 5th Ed. Pergamon, Oxford.

Cornbleet, S. (1976). *Microwave Optics*. Academic Press, New York.

Luneberg, R. K. (1964). *Mathematical Theory of Optics*. University of California Press, Berkeley, CA.

Morgan, S. P. (1958). General solution of the Luneburg lens problem. *J. Appl. Phys.* **29**, 1358–1368.

Appendix C
The Geometry of the Basic Compound Parabolic Concentrator

It is probably simple to obtain the basic properties of the CPC from the equation of the parabola in polar coordinates. Figure C.1 shows the coordinate system. The focal length f is the distance AF from the vertex to the focus. The equation of the parabola is, then,

$$r = \frac{2f}{1-\cos\phi} = \frac{f}{\sin^2\dfrac{\phi}{2}} \tag{C.1}$$

This result is given in elementary texts on coordinate geometry, and it may be verified by transforming to polars the more familiar cartesian form of $z = y^2/4f$ with axes as indicated in the figure.

We apply this to the design of the CPC as in Figure C.2. We first draw the entrance and exit apertures PP' and QQ' with the desired ratio in aperture between them, and we choose the distance between them so that an extreme ray PQ' (or $P'Q$) makes the maximum collecting angle θ_i with the concentrator axis. Then according to Section 4.6 the profile of the CPC between P' and Q' is a parabola with axis parallel to PQ' and with focus at Q, and this parabola can be expressed in terms of the polar coordinates (r, φ) as in the diagram.

For the exit aperture we have, using Equation (D.1),

$$QQ' = \frac{2f}{1-\cos(\pi/2 + \theta_i)}$$

so that, if $QQ' = 2a'$, in our usual notation

$$f = a'(1 + \sin\theta_i) \tag{C.2}$$

Next we find

$$QP' = \frac{2f}{1-\cos 2\theta_i}$$

or

$$QP' = \frac{a'(1 + \sin\theta_i)}{\sin^2\theta_i} \tag{C.3}$$

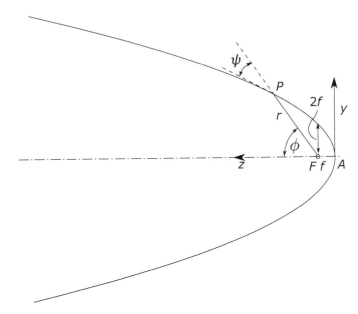

FIGURE C.1 The parabola in polar coordinates with origin at the focus, $r = 2f / (1 - \cos\phi) = f / \sin^2(1/2)\phi$.

Thus,

$$a + a' = QP' \sin\theta_i = a' \frac{(1 + \sin\theta_i)}{\sin\theta_i}$$

so that $a = a' / \sin\theta_i$, as required. Finally, for the length of the concentrator

$$L = QP' \cos\theta_i = a' \frac{(1 + \sin\theta_i)}{\tan\theta_i \sin\theta_i} = \frac{a + a'}{\tan\theta_i} \tag{C.4}$$

This, of course, is again as required by the way we set up the design of the CPC.

It is obvious from the geometry of the initial requirements for the CPC that the profile at P and P' should have its tangent parallel to the axis. This also can be verified from the preceding formulation. For the tangent of the angle between the curve and the radius vector is, in polars,

$$\tan\psi = r \frac{d\phi}{dr} = \frac{2f}{r \sin\phi}$$

for the parabola, and if we put $r = QP'$ and $\phi = 2\theta_i$, we find

$$\tan\psi = \tan\theta_i$$

as expected.

The parametric representation of the CPC profile given in Section 4.6 is easily obtained from Figure C.2 if we take an origin for cartesian coordinates at the

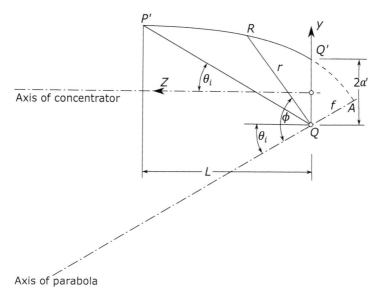

FIGURE C.2   Design of the CPC.

center of the exit aperture and z-axis along the concentrator axis. We have from the figure

$$y = r\sin(\phi - \theta_i) - a' = \frac{2f\sin(\phi - \theta_i)}{1 - \cos\phi} - a'$$

$$= \frac{2a'(1 + \sin\theta_i)\sin(\phi - \theta_i)}{1 - \cos\phi} - a'$$

$$z = r\cos(\phi - \theta_i) = \frac{2a'(1 + \sin\theta_i)\cos(\phi - \theta_i)}{1 - \cos\phi}$$

Appendix D
The θ_i/θ_o Concentrator

Figure D.1 shows the concentrator with appropriate notation. From the triangle $QQ'S$ we have

$$QS = 2a' \frac{\cos \dfrac{\theta_o - \theta_i}{2}}{\sin \dfrac{\theta_o + \theta_i}{2}} \tag{D.1}$$

Also, since the angle $R\hat{Q}S$ is $\pi - \theta_o - \theta_i$, the triangle RQS is isosceles, and so $QR = QS$.

From the polar equation of the parabola,

$$QR = \frac{2f}{1 - \cos \theta_o + \theta_i} \tag{D.2}$$

so from Equation (D.1) we find for the focal length of the parabola

$$f = 2a' \cos \frac{\theta_o - \theta_i}{2} \sin \frac{\theta_o + \theta_i}{2} = a' \sin \theta_i + \sin \theta_o \tag{D.3}$$

Again, using the polar equation of the parabola,

$$QP' = \frac{2f}{1 - \cos 2\theta_i} = a' \frac{\sin \theta_i + \sin \theta_o}{\sin^2 \theta_i}$$

and

$$a + a' = QP' \sin \theta_i = \frac{a'(\sin \theta_i + \sin \theta_o)}{\sin \theta_i} \tag{D.4}$$

From Equation (D.4) we find immediately

$$\frac{a}{a'} = \frac{\sin \theta_o}{\sin \theta_i} \tag{D.5}$$

so that the θ_i/θ_o concentrator is actually ideal in two dimensions.

The overall length of the concentrator is, from the figure,

$$L = (a + a') \cot \theta_i \tag{D.6}$$

just as for the basic CPC (Equation (D.4)).

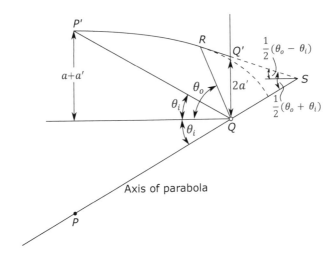

FIGURE D.1 Design of the θ_i/θ_o concentrator.

Appendix E
The Truncated Compound Parabolic Concentrator

In this appendix we give the derivations of the formulas in Section 5.7. We show in Figure E.1 the parabolic section in polar coordinates. For these coordinates we recall (see Appendix D) that the equation of the parabola is

$$r = \frac{2f}{1 - \cos\phi} = \frac{f}{\sin^2 \dfrac{\phi}{2}}$$

$$f = a'(1 + \sin\theta_i) \qquad\qquad\text{(E.1)}$$

If s is the arc length of the parabola, then

$$ds^2 = dr^2 + r^2 d\phi^2$$

so that

$$\frac{ds}{d\phi} = \sqrt{r^2 + \left(\frac{dr}{d\phi}\right)^2}$$

or

$$\frac{ds}{d\phi} = \frac{f}{\sin^3 \dfrac{\phi}{2}} \qquad\qquad\text{(E.2)}$$

from Equation (E.1). We have to integrate this to find the arc length, which is proportional to the reflector area in a 2D system. We have to evaluate the indefinite integral

$$s = f\int \frac{d\phi}{\sin^3 \dfrac{\phi}{2}}$$

On making the change of variable

$$\cosh u = \frac{1}{\sin \dfrac{\phi}{2}}$$

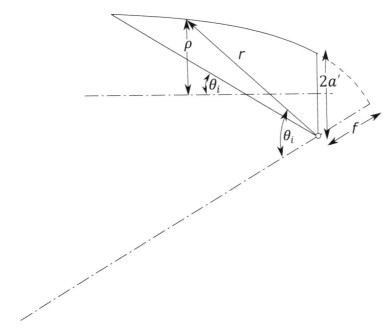

FIGURE E.1   Construction for the truncated CPC.

we obtain

$$s_T = -f \left\{ \frac{\cos \dfrac{\phi}{2}}{\sin^2 \dfrac{\phi}{2}} - \ln\left(\tan \dfrac{\phi}{4} \right) \right\} \Bigg|_{\phi_T}^{\theta_i + \pi/2} \qquad (E.3)$$

The radius of the entrance aperture is simply

$$a_T = r \sin(\phi - \theta_i) - a'$$

which gives Equation (5.18) immediately. Thus, Equation (5.22) for s_T/a_T is obtained. Likewise the length L_T of the truncated CPC is

$$L_T = r \cos(\phi - \theta_i)$$

from which Equation (5.19) follows immediately, and thence Equation (5.21).
For the 3D CPC the element of area of reflector is

$$dA = 2\pi \rho \, ds$$

where ρ is the radius of the CPC at the current point. Since ρ is $r \sin(\phi - \theta_i) - a'$ we have

$$dA = 2\pi \left\{ \frac{f \sin\left(\phi - \theta_i\right)}{\sin^2 \dfrac{\phi}{2}} - a' \right\} \frac{f}{\sin^3 \dfrac{\phi}{2}}\, d\phi$$

This can be integrated by making the same change of variable, $\cosh u = \operatorname{cosec}\left(\phi / 2\right)$, and the result for the ratio of area of collection surface divided by area of collecting aperture is

$$\frac{2f}{a_T^2} \left[f \sin\theta_i \left\{ \frac{3}{4} \ln \cot \frac{\phi}{4} + \frac{3}{4} \frac{\cos \dfrac{\phi}{2}}{\sin^2 \dfrac{\phi}{2}} + \frac{1}{2} \frac{\cos \dfrac{\phi}{2}}{\sin^4 \dfrac{\phi}{2}} \right\} \right.$$

$$\left. - \left(2f \sin\theta_i - a'\right) \left\{ \ln \cot \frac{\phi}{4} + \frac{\cos \dfrac{\phi}{2}}{\sin^2 \dfrac{\phi}{2}} \right\} - \frac{4f \cos\theta_i}{3\sin^3 \dfrac{\phi}{2}} \right]_{\phi_T}^{\theta_i + \pi/2}$$

(E.4)

Appendix F
Skew Rays in a Hyperboloidal Concentrator

We shall prove that any ray aimed at the rim of the entry aperture is reflected to some point on this rim. Let FF' in Figure F.1 be the diameter of the entry aperture, so that F and F' are the foci of the hyperbolas in the plane of the diagram, and let P be a point on a hyperbola. The bisector PR of angle FPF' is tangent to the hyperbola, and thus the tangent plane at P passes through PR. Any skew ray to the rim incident on the hyperbola at P passes through some point on the circle of which FF' is the diameter and that is in a plane perpendicular to the diagram, and all such rays generate an oblique circular cone. We now draw F_1F_1' through R to make the same angle with PR, as shown. A plane through F_1F_1' perpendicular to the diagram cuts the cone in a certain curve; the cone is a (degenerate) quadric surface, and therefore any plane intersects it in a conic section; but the plane through FF' intersects the cone in a circle by construction, and since the section through F_1F_1' has the same dimensions in and perpendicular to the plane of the diagram, it too must be a circle. An incident skew ray aimed at Q on the circle through FF' is reflected to some point Q' on the plane through FF', and by our construction this point is the mirror image in the tangent plane to the hyperboloid of the point in which the ray cuts the plane through F_1F_1'; thus Q' is on the circle through FF'.

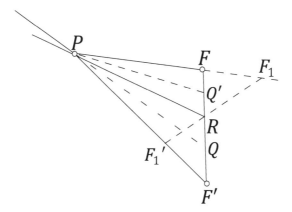

FIGURE F.1   Geometry of hyperboloid concentrator for skew rays.

Appendix G
Sine Relation for Hyperboloidal/ Lens Concentrator

With reference to Figure G.1 the lens images the input beams inclined at angles $\pm\theta$ to the axis onto the edges FF' of the virtual source. Therefore, by Fermat's principle $[QA'] + [A'F'] = [AF']$, where the brackets denote optical path length, and similarly for the other beam. From the geometry of the hyperbola, $[AF'] - [A'F'] = 2c$. Also $[QA'] = 2a \sin \theta$, since QA is normal to the beam. The desired result follows: $c = a \sin \theta$.

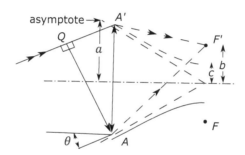

FIGURE G.1   The thin lens at AA' images the input beam inclined at angle θ to the axis onto F'.

Index

Printed and bound by CPI Group (UK) Ltd, Croydon, CR0 4YY

17/10/2024

01775682-0010